ChatGPT for Conversational AI and Chatbots

Learn how to automate conversations with the latest large language model technologies

Adrian Thompson

ChatGPT for Conversational AI and Chatbots

Group Product Manager: Niranjan Naikwadi
Publishing Product Manager: Nitin Nainani
Book Project Manager: Shambhavi Mishra
Senior Editor: Nazia Shaikh
Technical Editor: Rahul Limbachiya
Copy Editor: Safis Editing
Proofreader: Nazia Shaikh
Indexer: Tejal Soni
Production Designer: Aparna Bhagat
Senior DevRel Marketing Executive: Vinishka Kalra

First published: July 2024

Production reference: 1050724

Published by Packt Publishing Ltd.
Grosvenor House
11 St Paul's Square
Birmingham
B3 1RB, UK

ISBN 978-1-80512-953-0

www.packtpub.com

This book has been a long time in the making, and I am deeply grateful to my loving wife, Lucy, for her constant support and understanding of the late nights, dodged housework, and endless sacrifices she made to help me see it through to completion. To my kids, Fletcher and Amalie, your boundless energy and smiles have been my greatest motivation.

– Adrian Thompson

Contributors

About the author

Adrian Thompson, a leading expert in **Conversational AI (CAI)**, has been at the forefront of building chatbots and voice assistants since 2018. As the founder of The Bot Forge, he established a pioneering consultancy service recognized for its cutting-edge innovations in CAI. Adrian's expertise spans the latest **large language models (LLMs)**, such as ChatGPT and Google Gemini. A versatile programmer and prompt engineer, he is adept in the LLM landscape and related technologies. At loveholidays, Adrian enhances customer experiences by developing advanced CAI systems. He is passionate about the latest advancements in CAI and is renowned for his proficiency in conversation design, AI development, and strategic consultancy across various industries.

I would like to first and foremost thank my loving and patient wife, kids, friends, and family for their continued support, patience, and encouragement throughout the long process of writing this book. Thanks also to the Conversational AI team at loveholidays for listening to my ideas and the many interesting people I spoke to during my days building up The Bot Forge who provided inspiration and encouragement.

About the reviewers

Krishnan Raghavan is an IT professional with over 20 years of experience in the areas of software development and delivery excellence across multiple domains and technologies, ranging from C++ to Java, Python, Angular, Golang, and data warehousing.

When not working, Krishnan likes to spend time with his wife and daughter, as well as reading fiction and nonfiction as well as technical books and participating in hackathons. Krishnan tries to give back to the community by being part of the GDG – Pune Volunteer Group.

You can connect with Krishnan at `mailtokrishnan@gmail.com` or via LinkedIn: `www.linkedin.com/in/krishnan-Raghavan`.

I would like to thank my wife, Anita, and daughter, Ananya, for giving me the time and space to review this book.

Manoj Palaniswamy is a senior technical staff member at Kyndryl and plays a technical leadership role within the application, data, and AI practice. He has been working in the IT industry for more than 17 years, and his domain expertise includes enterprise AI strategy, data management, AIOps, MLOps, AI platform engineering, and hybrid cloud IT infrastructure and analytics. He has led many cross-cultural technical teams across the globe on complex technical projects and closely works with clients to translate their business requirements into technical solutions. Manoj holds two patents in the area of machine learning and workload optimization on VMs.

Table of Contents

2

Using ChatGPT with Conversation Design   31

Part 2: Using ChatGPT, Prompt Engineering, and Exploring LangChain

3

ChatGPT Mastery – Unlocking Its Full Potential   53

4

Prompt Engineering with ChatGPT   75

5

Getting Started with LangChain   97

6

Part 3: Building and Enhancing ChatGPT-Powered Applications

7

8

Creating Your Own LangChain Chatbot Example   167

9

The Future of Conversational AI with LLMs   189

Preface

Hello there!

Since its release in 2022, ChatGPT has taken the world by storm, ushering in the era of the **Large Language Model (LLM)**, and revolutionizing the field of conversational AI.

This book, *ChatGPT for Chatbots and Conversational AI: Leveraging advanced language models to revolutionize conversational experiences*, is your comprehensive guide to understanding and mastering the latest advancements in conversational AI, with a special focus on OpenAI's ChatGPT.

The advent of ChatGPT has marked a significant leap in the capabilities of chatbots and virtual assistants. Unlike traditional rule-based systems, LLMs generate human-like text based on the context provided. This allows for more fluid, natural, and contextually appropriate interactions, transforming the user experience in various applications from customer service to personal assistance.

In this book, we will explore how ChatGPT and LLMs are reshaping the landscape of conversational AI.

The content is structured to provide both theoretical insights and practical applications. We start with the basics, covering the evolution of chatbots and LLMs and their use in conversation design, as well as a solid introduction to ChatGPT and OpenAI's models. We then move on to more advanced topics, including the latest techniques in prompt engineering, a deep dive into LangChain, the use of **retrieval-augmented generation (RAG)** systems, and the creation of your own sophisticated chatbot application.

There are still many thousands of conventional **natural language understanding (NLU)** powered systems in production and I know there are still many questions as to when, how, or whether you should transition to LLM-powered systems, which I'll try to address by highlighting the challenges you may encounter and the strategies to overcome them.

Bringing your LLM applications into production involves more than just technical implementation. It requires robust evaluation strategies to ensure the models perform as expected in real-world scenarios. This book emphasizes the importance of continuous monitoring, rigorous testing, and iterative improvements to maintain high-quality interactions and user satisfaction.

Throughout the book, we emphasize real-world applications and use cases, ensuring that the concepts you learn are directly applicable to your projects. By the end of this journey, you will not only understand how to implement ChatGPT in various scenarios but also be prepared for future trends in the field.

This book stands out by providing a comprehensive overview of the ChatGPT landscape, combining the latest research with practical insights. We delve into foundational concepts, advanced techniques, innovative tools, and industry best practices, offering a clear roadmap for anyone aiming to develop sophisticated conversational applications using ChatGPT.

There is a lot to cover and it's a rapidly changing landscape. Join me to unlock ChatGPT's full potential for your conversational AI applications.

Happy reading!

Who this book is for

This book is designed for individuals who are eager to harness the power of ChatGPT and other advanced language models in their conversational AI projects. The primary audiences are as follows:

- **Conversational AI professionals**: From conversation designers to customer experience managers and conversational AI leads, anyone involved in working with chatbots, voice assistants, smart IVR systems, and multimodel customer support will find practical insights into leveraging ChatGPT to create more engaging and natural interactions

- **Developers and conversational AI engineers**: Those with a technical background looking to transition from conventional NLU systems to LLM-powered applications will benefit from detailed guidance on integration, deployment, and overcoming common challenges

- **Product owners**: Leaders overseeing the development of AI-driven products will gain a comprehensive understanding of the latest technologies, tools, and best practices necessary to bring innovative conversational solutions to market

- **AI enthusiasts and researchers**: Individuals passionate about the advancements in AI will appreciate the in-depth exploration of ChatGPT, prompt engineering, LangChain, and other innovative technologies that are shaping the future of conversational AI

Whether you are looking to improve existing chatbot applications, start new AI projects, or stay ahead of industry trends, this book provides the knowledge and tools to help you succeed.

What this book covers

Chapter 1, An Introduction to Chatbots, Conversational AI, and ChatGPT, introduces the fundamentals of chatbots and conversational AI. It explores their evolution, various types, and impact across industries such as e-commerce, customer service, and healthcare. The chapter delves into OpenAI's ChatGPT, detailing its development, capabilities, and limitations, and examines how it fits into the broader conversational AI landscape.

Chapter 2, Using ChatGPT with Conversation Design, delves into the role of conversation designers and the impact of LLMs such as ChatGPT on conversation design. It covers practical applications, including simulating conversations and creating personas, and emphasizes the importance of testing and iteration in developing engaging conversational AI systems. By the end, you'll understand how to effectively use ChatGPT in conversation design.

Chapter 3, ChatGPT Mastery – Unlocking Its Full Potential, explores the technical aspects of interacting with ChatGPT, covering the webchat interface, OpenAI Playground, API usage, and official libraries. You'll learn the differences between the Free and Plus versions, custom instructions, and how to use the playground. By the end, you'll be equipped to choose the best interaction method for your needs.

Chapter 4, Prompt Engineering with ChatGPT, focuses on mastering prompt engineering. You'll learn about the core components of successful prompts, strategies for tone and complexity, and techniques for enhancing readability. This chapter equips you to create prompts that maximize the capabilities of ChatGPT, ensuring precise and relevant interactions.

Chapter 5, Getting Started with LangChain, introduces LangChain, an open-source framework for building complex LLM applications. You'll learn about the core components, the **LangChain Expression Language** (**LCEL**), and various types of chains you can create. By the end, you'll have a solid foundation to engineer LangChain applications and tackle more advanced functions in the next chapter.

Chapter 6, Advanced Debugging, Monitoring, and Retrieval with LangChain, Agents, and Tools, delves into advanced LangChain topics, focusing on debugging techniques, leveraging agents and tools, and understanding memory for LLM-powered conversational experiences. You'll explore the LangSmith platform, out-of-the-box tools, and custom tool creation for agents. This chapter builds on previous concepts, enabling you to develop more complex LangChain applications.

Chapter 7, Vector Stores as Knowledge Bases for Retrieval-augmented Generation, introduces RAG, a popular use case for LLMs. You'll learn about the essential steps to create a RAG system and how to implement these processes using LangChain. Through a real-world example, you'll gain a solid foundation in mastering RAG concepts and techniques.

Chapter 8, Creating Your Own LangChain Chatbot Example, brings together key concepts from previous chapters into a practical project. You'll build a ChatGPT-powered chatbot capable of answering questions about your data and handling complex tasks. We'll look at scoping your project, preparing your data for RAG, creating agent tools, and using them in LangChain, and finally, bringing it all together with the Streamlit framework to create your own Chatbot UI. By the end, you'll have a functional chatbot and a solid understanding of crafting sophisticated conversational agents with ChatGPT.

Chapter 9, The Future of Conversational AI with LLMs, delves into taking ChatGPT applications to production, examining lessons learned in the industry, and exploring strategies for success. It explores alternatives to ChatGPT, particularly smaller language models, and discusses future trends in LLMs. By the end, you'll be equipped to navigate the evolving conversational AI landscape and plan for your organization's future.

To get the most out of this book

Familiarity with the fundamental concepts of chatbots and conversational AI will be beneficial, though not essential. Some expertise in coding, particularly in Python, is recommended to follow the examples and exercises effectively. A working Python environment running Python 3.10 is required. We suggest using Anaconda to set this up, with Jupyter Notebook installed, as all chapter examples use Jupyter Notebook.

Software/hardware covered in the book	Operating system requirements
Python 3.10	Cross-platform
LangChain	Cross-platform

A free OpenAI account is required so you can get an OpenAI API key for accessing ChatGPT and other related services: https://openai.com/

A LangSmith account is required so you can create an API key for monitoring and testing your LangChain applications: https://smith.langchain.com

By ensuring you meet these prerequisites, you'll be well prepared to dive into the practical applications covered in this book.

If you are using the digital version of this book, we advise you to type the code yourself or access the code from the book's GitHub repository (a link is available in the next section). Doing so will help you avoid any potential errors related to the copying and pasting of code.

Download the example code files

You can download the example code files for this book from GitHub at https://github.com/PacktPublishing/ChatGPT-for-Conversational-AI-and-Chatbots. If there's an update to the code, it will be updated in the GitHub repository.

We also have other code bundles from our rich catalog of books and videos available at https://github.com/PacktPublishing/. Check them out!

Conventions used

There are a number of text conventions used throughout this book.

`Code in text`: Indicates code words in text, database table names, folder names, filenames, file extensions, pathnames, dummy URLs, user input, and Twitter handles. Here is an example: "Make sure you have installed LangSmith by running pip install -U langsmith."

A block of code is set as follows:

```
def format_evaluator_inputs(run: Run, example: Example):
    return {
        "input": example.inputs["question"],
        "prediction": next(iter(run.outputs.values())),
        "reference": example.outputs["answer"],
    }
```

Any command-line input or output is written as follows:

```
splitter = RecursiveCharacterTextSplitter(chunk_size=250,
    chunk_overlap=20)
```

Bold: Indicates a new term, an important word, or words that you see onscreen. For instance, words in menus or dialog boxes appear in **bold**. Here is an example: "Clicking on the SUCCESS or FAILURE button will display details of the test for each input and output in the dataset."

> **Tips or important notes**
> Appear like this.

Get in touch

Feedback from our readers is always welcome.

General feedback: If you have questions about any aspect of this book, email us at customercare@ packtpub.com and mention the book title in the subject of your message.

Errata: Although we have taken every care to ensure the accuracy of our content, mistakes do happen. If you have found a mistake in this book, we would be grateful if you would report this to us. Please visit www.packtpub.com/support/errata and fill in the form.

Piracy: If you come across any illegal copies of our works in any form on the internet, we would be grateful if you would provide us with the location address or website name. Please contact us at copyright@packt.com with a link to the material.

If you are interested in becoming an author: If there is a topic that you have expertise in and you are interested in either writing or contributing to a book, please visit authors.packtpub.com.

Share Your Thoughts

Once you've read *ChatGPT for Conversational AI and Chatbots*, we'd love to hear your thoughts! Scan the QR code below to go straight to the Amazon review page for this book and share your feedback.

https://packt.link/r/1-805-12953-8

Your review is important to us and the tech community and will help us make sure we're delivering excellent quality content.

Download a free PDF copy of this book

Thanks for purchasing this book!

Do you like to read on the go but are unable to carry your print books everywhere?

Is your eBook purchase not compatible with the device of your choice?

Don't worry, now with every Packt book you get a DRM-free PDF version of that book at no cost.

Read anywhere, any place, on any device. Search, copy, and paste code from your favorite technical books directly into your application.

The perks don't stop there, you can get exclusive access to discounts, newsletters, and great free content in your inbox daily

Follow these simple steps to get the benefits:

1. Scan the QR code or visit the link below

https://packt.link/free-ebook/978-1-80512-953-0

2. Submit your proof of purchase
3. That's it! We'll send your free PDF and other benefits to your email directly

Part 1:
Foundations of
Conversational AI

In this part, you will gain a comprehensive understanding of the foundations of conversational AI. We begin by exploring the fundamentals of chatbots, their evolution, and their impact across various industries. You will also learn about the revolutionary capabilities of OpenAI's ChatGPT, its applications, and its limitations. Additionally, we will delve into how ChatGPT can be used in conversation design, enhancing user interactions and creating more natural conversational experiences.

This part has the following chapters:

- *Chapter 1, An Introduction to Chatbots, Conversational AI, and ChatGPT*
- *Chapter 2, Using ChatGPT with Conversation Design*

1

An Introduction to Chatbots, Conversational AI, and ChatGPT

Welcome to the exciting world of chatbots, **conversational artificial intelligence (conversational AI)**, and ChatGPT!

In this introductory chapter, we introduce you to chatbots and conversational AI.

Our journey will begin by exploring the fundamentals of chatbots and their varying types, tracing the evolution of these digital conversationalists, and assessing their impact across multiple industries, such as e-commerce, customer service, and healthcare.

We'll then look at how the conversational AI landscape has changed over time and the recent rapid rise of **large language model (LLM)** technologies, specifically the rise of OpenAI's ChatGPT, **Generative Pre-trained Transformer 3 (GPT-3)**, and GPT-4 models.

While journeying through this chapter, you'll gain a comprehensive understanding of OpenAI's ChatGPT. We'll pull back the curtain on this marvel of technology to examine its development, capabilities, and limitations. You'll see firsthand how ChatGPT fits into the broader conversational AI landscape and how it has revolutionized various sectors with its diverse applications.

What makes this chapter valuable? Our aim is to look at the broader picture of conversational AI and gain a solid understanding of ChatGPT, essentially laying the groundwork for the practical skills that you'll learn later in this book. By the end of this chapter, you'll have a rich understanding of chatbots and conversational AI, particularly the real-world applications and potential of ChatGPT and the opportunities ChatGPT offers in comparison to earlier conversational AI technologies. This insight will enable you to see the potential for ChatGPT's conversational AI applications to bring real benefit to your organization, enhancing user experiences and business processes and bringing a novel edge to your industry. Let's embark on this enlightening journey together!

In this chapter, we're going to cover the following main topics:

- What are chatbots and conversational AI?
- Evolution of chatbots and conversational AI
- Understanding conversational AI applications
- What is OpenAI's ChatGPT?
- Capabilities and applications of ChatGPT
- Limitations of ChatGPT

What are chatbots and conversational AI?

When we think of conversational AI, we reference an evolving field of AI that equips computer systems with the ability to communicate with humans using **natural language** (**NL**) in an interactive manner.

Over the past decade, conversational AI-powered voice- and text-based assistants have become embedded in our daily lives, enhancing user experience across diverse platforms and tackling a wide range of use cases. Revolutionizing the way businesses interact with their customers, they have become critical components of a modern digital strategy.

This section aims to provide a comprehensive overview of the field of conversational AI. It will delve into the history of chatbots and conversational AI, highlighting key milestones that have shaped the progression from rudimentary rule-based systems to sophisticated AI-driven chatbots. We will examine advancements in technology, such as **NL understanding** (**NLU**) and the progression from rule-based chatbots to AI-powered chatbots, emphasizing their capabilities and limitations. Overall, this section aims to provide readers with a clear understanding of the significance and potential of conversational AI in revolutionizing interactions between humans and computers.

A brief history of conversational AI

The first chatbots were created in the mid-20th century with the first recorded example, ELIZA, emerging from the hallowed halls of **Massachusetts Institute of Technology** (**MIT**) in 1966, thanks to Joseph Weizenbaum's pioneering work. ELIZA was architected to **mirror** the linguistic patterns of a psychotherapist, capable of sustaining straightforward interactions with human users. This groundbreaking creation established the fundamental bedrock for the later development of more complex conversational AI systems.

Originally, these elementary chatbots could execute only a restricted set of pre-programmed responses or decision trees. However, they have since undergone a remarkable evolution, broadening their capability to respond to a diverse range of inputs.

The advent of revolutionary technologies such as **machine learning (ML)**, **NL processing (NLP)**, and NLU fostered the transformation of rudimentary chatbots into intricate conversational AI systems. These advanced systems not only comprehend and reciprocate human speech in a more organic, instinctive manner but can also engage in more dynamic conversations. A pivotal feature of these modern systems is their ability to learn and adapt over time using ML, progressively improving their efficiency and acceptance among users and, therefore, their demand.

The development and evolution of conversational AI has been marked by significant milestones that highlight its steady progression from rudimentary chatbots to advanced AI systems that can simulate human-like conversation.

Each of these milestones represents a significant advancement in the development of conversational AI, paving the way for the sophisticated intent-based systems we see today. However, it's not an exaggeration to state that it wasn't until November 2022 that conversational AI really took off.

To emphasize the phenomenal growth and growth of ChatGPT, consider the fact that it amassed over a million users within the span of just 5 days post-launch in November 2022. This growth trajectory outpaced even those of tech giants such as Netflix, Facebook, Instagram, and Zoom.

An overview of chatbots and automated assistants

What exactly do we mean when we speak of chatbots or digital assistants? The terminology of *chatbot* has been subject to debate, and some argue that it bears the brunt of past disappointments when the technology was still in its nascent stages. For the purpose of this book, whether we refer to them as chatbots, digital assistants, or any other term, we are essentially addressing a computer program that communicates with people, either through text or voice. This interactive communication can take place across various mediums, such as websites, mobile apps, messaging platforms, collaborative tools, smart speakers, digital human avatars, or **interactive voice response (IVR)** systems.

To further delineate, we can categorize these entities into two main types of conversational agents that have different capabilities:

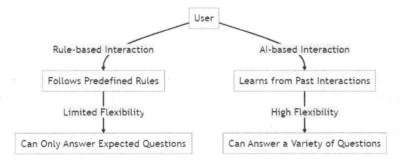

Figure 1.1 – Comparison between rule-based and AI-powered chatbots

Rule-based chatbots/digital assistants

Simple, rule-based chatbots operate based on a set of predefined rules or FAQs. They're designed to answer frequently asked questions, provide information, or guide users through specific tasks. However, their capacity to handle interactions is confined to pre-programmed knowledge, and they can struggle with requests that deviate from their predefined scripts:

Figure 1.2 – An interaction with a rule-based chatbot

In short, these chatbots can't understand any language outside the commands they're programmed to understand. These types of chatbots don't utilize any kind of ML technology and are limited and brittle in their use case and functionality. If a user makes a request that is not within the bounds of their conversational capability or the decision tree, then they will often fall over, resulting in a poor user experience.

Conversational AI-driven chatbots/digital assistants

Conversational AI refers to the technology that enables computers or machines to engage in natural and human-like conversations with users. It combines various fields, such as NLP, NLU, ML, and dialogue management, to understand and respond to user inputs in a conversational manner. Nowadays conversational AI is the common term for a field that encompasses chatbots and voice assistants and the systems and technologies used to create them.

As technology has advanced, conversational AI has evolved to become more sophisticated, capable of understanding complex queries, providing meaningful and contextually relevant responses, and offering a much more robust and far more dynamic and adaptable approach.

The core features of a conversational AI chatbot are made up of the following:

- **Understanding of context**: One of the most significant features of NLU is its ability to comprehend the context of a conversation. This means it doesn't just process words in isolation but also considers the context in which those words are used. This allows the AI to correctly interpret the meaning behind a user's input, even if it's ambiguous or complex.

- **Semantic understanding**: NLU systems have a high degree of semantic understanding. They can grasp the meaning behind a user's words, including synonyms, slang, regional dialects, and industry-specific jargon, by using entity recognition.

- **Handling of ambiguity**: In human communication, ambiguity is common. We often use vague or unclear expressions that can be interpreted in different ways. An NLU system can handle such ambiguity by using contextual clues to infer the intended meaning. For example, consider the sentence, "*I saw her duck.*" This sentence can be interpreted in multiple ways: it could mean that the speaker saw a woman dodge to avoid something, or it could mean that the speaker saw a duck that belongs to a woman. If the previous conversation was about avoiding flying objects, the system would infer the first meaning. If the conversation was about pets or animals, it would infer the second meaning.

- **Intent recognition**: NLU systems can identify the user's intent behind their words. This means they don't just process what the user is saying but also understand what the user wants to achieve. This is key to providing a helpful and relevant response.

- **Entity recognition**: Entity recognition is a key technique in NLP. It's the task of identifying and classifying named entities (such as persons, organizations, locations, product types, dates, and other variables) within the text of a provided input.

- **Slot filling**: Slot filling is the functionality to extract information from inputs based on entity type. Relying heavily on entity recognition, it's about identifying and extracting specific pieces of information from an input and placing them into predefined **slots** in memory for later use. For example, in a hotel booking dialogue system, slots might include type of room, time of booking, and number of people.

- **Sentiment analysis**: Many NLU systems can analyze the sentiment behind a user's words. This allows them to recognize when a user is happy, frustrated, angry, or sad and respond in a way that is sensitive to the user's emotional state.

- **Conversation management**: NLU systems can manage complex conversations. They can keep track of a conversation's history and use this information to inform their responses. They can handle interruptions, follow-up questions, and changes in the conversation topic, providing a more human-like interaction.

- **Multilingual capability**: Advanced NLU systems can understand and communicate in multiple languages. This allows businesses to provide customer service and support to a global audience.

- **Integration with external systems**: NLU systems can often be integrated with other systems, such as **customer relationship management** (**CRM**) systems, databases, or other APIs. This allows the AI to retrieve and utilize real-time data when interacting with the user, providing more personalized and relevant responses.

The following diagram shows the steps for an interaction with a modern conversational AI chatbot:

Figure 1.3 – An interaction with an AI-powered chatbot

Over the past 5 years, there has been a huge growth in conversational AI platforms, with technology vendors large and small striving to build the best low-code, no-code, pro-code solutions. Everyone who has worked in the field of conversational AI over the years has an opinion on what is the best or their favorite platform to build conversational experiences.

The key component of all these systems is that they are all intent-based. The AI technology handles the user intent based on a set of trained intents and provides an answer based on its understanding of the context of the conversation.

Conversational AI platforms

Modern conversational AI systems employ several features to manage a conversation. These features are commonly the following:

- The ability to understand NL input from users via text or voice
- The ability to generate NL output that is relevant, coherent, and engaging
- The ability to understand multiple intents, entities, and contexts in a conversation
- The ability to learn from user feedback and data to improve their accuracy and understanding over time
- The ability to provide a consistent and personalized experience across multiple platforms and channels

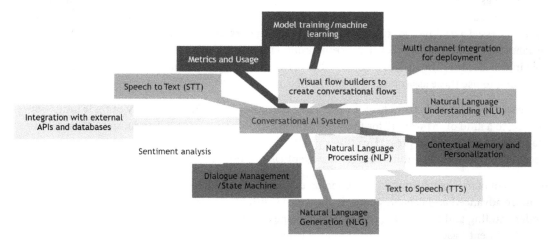

Figure 1.4 – The elements of a modern conversational AI system

There are many great conversational AI systems on the market, with vendors ranging from the biggest tech providers to tech start-ups concentrating solely on conversational AI solutions.

In reality, it's not always the end conversational experience created by one of these systems that makes for a successful implementation but the features of the system in place and the usability of these features that enable users to create conversational experiences.

To appreciate today's advanced conversational AI, let's trace the evolution of chatbots from their humble beginnings.

Evolution of chatbots and conversational AI

When chatbots first entered the scene, they were hailed as a groundbreaking evolution in technology, poised to revolutionize the way businesses communicate with their customers.

The early versions, however, were not as impressive as many had hoped. Coupled with the growth of chatbot platforms and the ease with which anyone could create a chatbot, this led to a bad user experience and an overall lackluster performance, which hampered their reputation.

One of the primary failings of early chatbots was their lack of sophistication and nuance in understanding and responding to human language. These rudimentary versions were generally rule-based, relying heavily on pre-programmed scripts. This meant they were only equipped to handle very specific inputs and could easily be thrown off by anything that deviated from their script. As a result, users often found themselves in frustratingly circular conversations, where the chatbot failed to comprehend their queries and kept providing irrelevant or nonsensical responses.

The voice assistants' journey has also been marked by highs and lows. While they enjoyed initial popularity, particularly with the advent of Amazon's Alexa, Apple's Siri, and Google Assistant, they have witnessed a decline over time. The accuracy and understanding of these voice assistants have led to some user frustration. Moreover, concerns around privacy and data security have played a significant role in this decline. The idea of a device "always listening" has made many users uncomfortable and unwilling to use these assistants.

The early hype around chatbots and voice assistants contrasted with the reality of their limitations, leading to a kind of "AI winter" for these technologies. The media was quick to jump on these failures, leading to a wave of bad press. Headlines proclaiming the "death" of chatbots, and voice assistants became commonplace, leading to a general skepticism about the potential of these technologies.

The advent of more sophisticated AI technologies such as NLP and ML has allowed for the development of more advanced and capable chatbots and voice assistants. These newer iterations are far better at understanding and responding to human language and can learn and adapt over time. In the same way that intent-based conversational AI has superseded rule-based agents, **LLMs** look to do the same.

LLMs have been in development for several years. However, it's only in the last couple of years with the emergence of GPT-3 technologies from OpenAI that the technology has started to get some exposure and enter into the mainstream. But in November 2022, all this changed with the explosion of ChatGPT. This has sparked a technological race, with hundreds of organizations striving to create the most advanced LLM. With ChatGPT and now GPT-4 leading the charge, they are arguably the ones to beat. In fact, it's not just the LLM technologies that are advancing but also other associated technologies that make up part of the ecosystem.

Conversational AI professionals can see that there are opportunities with this huge leap in AI-powered communication. GPT models trained for chat can make conversations with chatbots and automated assistants seem incredibly human-like. In the next section, we'll look at these applications in more detail.

Understanding conversational AI applications

Before looking at ChatGPT, it's important to consider use cases for conversational AI applications and initially consider them in the context of what's been possible with traditional conversational AI systems and technology.

Conversational AI has emerged as a powerful tool in numerous industries, from customer service to healthcare, banking, insurance, retail, and human resources. By automating routine tasks and enabling enhanced user interactions, conversational AI applications have changed the landscape of many domains. Let's look at some existing examples of conversational AI.

Customer service

Conversational AI has been hugely successful in the customer support use case, streamlining processes and improving user experience in myriad ways. By adopting AI-powered chatbots and virtual assistants, businesses can handle large volumes of customer interactions seamlessly and efficiently.

Conversational AI offers several capabilities and improvements in this use case. First, agents can interact with multiple customers simultaneously, resolving queries instantaneously. This eliminates long waiting times, a common source of customer dissatisfaction in traditional support systems.

Second, the round-the-clock availability of AI support greatly enhances customer experience. Conversational AI can cater to customer inquiries at any hour, including holidays, ensuring continuous support.

Third, the integration of AI with CRM systems and other databases provides personalized experiences. Conversational AI can use customer data to understand their preferences and purchasing history and deliver customized solutions or product recommendations, boosting customer satisfaction and potentially driving sales.

Lastly, conversational AI can help businesses save resources. AI can handle routine, repetitive tasks, freeing up human agents to tackle more complex and sensitive issues. This not only improves efficiency but also reduces operational costs.

loveholidays, a UK-based online travel agency, implemented a chatbot named Sandy. Sandy was designed to respond to customer queries in real time, reducing wait times and enhancing the customer experience. This bot responds to a high volume of interactions around the clock, resolving simple issues more efficiently than human representatives. This is a typical example of how conversational AI is revolutionizing customer service, from answering routine questions to proactive customer engagement.

Language translation

Language translation systems, such as Google Translate and DeepL, use conversational AI to facilitate real-time translation of spoken or written communication. Consider a use case where a French-speaking user needs to communicate with a Chinese associate. Using Google Translate, they can communicate seamlessly, each in their native language, thus breaking language barriers and enabling smooth communication.

Education

AI-based educational platforms such as Carnegie Learning leverage AI to provide personalized learning experiences. These platforms adapt to individual student needs, delivering tailored instructions and feedback. By studying student habits, these algorithms personalize their learning experience, offering a more effective, personalized learning approach.

Healthcare

In healthcare, conversational AI is employed for patient triage and preliminary diagnosis. Babylon Health, for example, has developed a chatbot that conducts an initial symptom check, recommending further medical attention or home care based on the findings. This bot can also schedule appointments and refill prescriptions, significantly enhancing patient care and satisfaction.

Other use-case examples include those created for mental health applications; for example, Woebot is an AI-powered personal mental health companion designed to support users in maintaining their emotional well-being. Utilizing conversational AI with NLP capabilities, it gradually learns about the user's mental state, offering tailored, clinically proven strategies based on **cognitive behavioral therapy** (**CBT**), **interpersonal psychotherapy** (**IPT**), and **dialectical behavioral therapy** (**DBT**). This innovative combination of sophisticated AI and proven therapeutic strategies creates a uniquely accessible mental health chatbot that uses smart conversational abilities and a slick chat **user interface** (**UI**).

Banking and insurance

The banking and insurance sectors use conversational AI for routine inquiries, account management, bill payment, claims processing, and policy management. Some establishments, such as JPMorgan Chase, use conversational AI to provide personalized financial advice and recommendations, creating a more engaging customer experience.

Retail

In retail, chatbots enhance customer service by recommending products, tracking orders, and answering routine queries. An AI-powered bot, for example, could recommend products based on a customer's browsing history, creating a personalized shopping experience. They also aid in tailoring marketing efforts, contributing to increased sales and customer retention.

Human resources

In human resources, AI-driven chatbots are making recruitment tasks more efficient. Mya, for instance, answers FAQs, screens candidates, schedules interviews, and manages resumes. Mya reported effective engagement with 92% of their candidates during the first 10,000 conversations, significantly reducing recruiters' workload.

In conclusion, the applications of conversational AI are vast and continually evolving. Whether it's enabling more efficient customer service, personalizing education, streamlining healthcare, revolutionizing banking, insurance, and retail, or enhancing HR processes, conversational AI is set to continue its transformational role across various industries.

While the journey has been bumpy, the story of chatbots and voice assistants is far from over. As these technologies continue to evolve, they have the potential to truly transform the way we communicate with machines and with each other. As with any technological innovation, the key is to learn from past mistakes and continue moving forward.

Conversational AI as a training tool – the emergence of digital humans

As the capabilities of conversational AI continue to evolve, we are witnessing the rise of "digital humans," which are hyper-realistic, AI-powered virtual beings that interact with users in human-like ways. These digital humans, coupled with the power of conversational AI, are creating new frontiers in health, education, training, and customer service for many organizations; for example, Deutsche Telekom, Vodafone, Kiehl, and UBS.

Companies such as UneeQ and Soul Machines are at the forefront of developing digital humans that can express emotions, adapt their responses based on the context, and learn from each interaction. These types of innovative interfaces integrate directly with modern conversational AI systems, handling **speech-to-text** (**STT**) and **text-to-speech** (**TTS**) capabilities and supporting **Speech Synthesis Markup Language** (**SSML**) and their own action programming languages to show human emotions and gestures.

In the realm of education and training, digital humans powered by conversational AI can be used to create immersive learning experiences that are personalized to each learner's needs. For instance, a digital human can act as a tutor, providing one-on-one tutoring in various subjects, or as a training assistant in corporate settings, guiding employees through complex processes or systems.

Digital human vendors have been early adopters of LLMs and **generative AI** (**GenAI**). The possibilities are truly fascinating and open a wide array of opportunities for further exploration and development in the field of AI. Many companies have realized that the combination of advanced digital human and conversational services powered by LLMs enables unique brand experiences.

Creating agents in digital human form helps to create a more personal, empathetic interaction and allows applications to be far wider-ranging. Some notable examples are the following:

- **Mental health applications**: Ellie, a digital human developed by the University of Southern California' Institute for Creative Technologies, has been used to conduct interviews with people to screen for signs of depression and post-traumatic stress disorder.

- **Medical education**: Digital patients can be used to train medical students in patient interaction and diagnosis.

- Digital trainers, creating immersive learning experiences that are personalized to each learner's needs. Think of your own personal tutor or a training assistant in corporate settings, guiding employees through complex processes or systems.

In the past, even the most advanced intent-based conversational AI systems struggled to provide the breadth of conversational experience needed to support these types of agents, particularly when conversations involved multiple turns and complex contexts. LLM technology is much more capable of offering realistic dialogue.

Conclusion

We've taken a comprehensive look at the realm of conversational AI, exploring its applications and acknowledging its limitations. Now, let's shift our focus to the enhanced abilities of ChatGPT. This powerful tool has the potential to bring about a fundamental shift in the conversational AI sector, paving the way for the creation of more sophisticated virtual assistants, chatbots, and various other communication systems and devices.

What is OpenAI's ChatGPT?

ChatGPT is OpenAI's chat-specific LLM that uses ML to generate human-like responses to your text inputs.

Within just over a month of launching in November 2022, ChatGPT had over 100 million users. By February 2023, its total website visits had skyrocketed to a billion. This meteoric rise has made ChatGPT the fastest-growing consumer software application in history.

ChatGPT can explain scientific and technical concepts in a desired style or programming language, brainstorm basically anything you can think of asking... and yes, of course, hold complex conversations!

Understanding the large language technology behind ChatGPT

ChatGPT is built on a type of AI called an LLM. Within the domain of LLMs, there are two primary types: basic LLMs and instruction-optimized LLMs. Understanding the difference between these two types is important when examining OpenAI's ChatGPT.

Exploring basic LLMs

Basic LLMs are programmed to predict the subsequent word in a string of words, using a vast text dataset obtained from diverse online sources. Their main function is to forecast the most probable word that would succeed a given sequence of words.

At this point, it's not actually words but tokens that are being predicted; more on that later in this section.

To illustrate the context of prediction, let's look at some examples, starting with a narrative generation. If we feed the phrase *Hey diddle diddle, the cat and the fiddle* to a basic LLM, it might continue with *the cow jumped over the moon*. It uses learned patterns from its vast corpus to predict an interesting and contextually relevant next token based on its text training data.

Here's an example of a specific query not being perfectly answered: if you were to ask *Who was the president of the United States in 1988?* the basic LLM, instead of providing the direct answer (Ronald Reagan), might return with a statement such as *The president of the United States is elected every 4 years during the general election*. This response, while related to the topic of US presidents, doesn't answer the original question directly, a potential downside of the base LLM's probabilistic prediction approach.

Understanding instruction-optimized LLMs and reinforcement learning from human feedback

In contrast to base LLMs, instruction-optimized LLMs are specifically developed to adhere to instructions more accurately. As such, they are the main area of interest in current LLM research and applications.

Let's look at the process, which can be broken down into three steps:

1. Pretraining a **language model (LM)**
2. Gathering data and training a reward model
3. Fine-tuning the LM with **reinforcement learning (RL)**

To start the process, an instruction-optimized LLM is initially trained on the same large text dataset as a basic LLM. OpenAI used a smaller version of GPT-3 for its first popular **RL from human feedback (RLHF)** model, InstructGPT. This model undergoes more fine-tuning with specific instruction-following data, leading to more precise responses.

Hence, when asked the question *Who won the Pulitzer Prize for fiction in 2020?* an instruction-optimized LLM would likely respond directly with: *The Pulitzer Prize for fiction for 2020 was awarded to The Nickel Boys by Colson Whitehead.*

To improve the quality of the LLM's output, a common process is then to obtain human ratings of the quality of many different LLM outputs on certain criteria; for example, if the output is helpful, truthful, and harmless.

This instruction learning is further honed using a method known as RLHF, which essentially optimizes the LM to be more useful. RLHF is a pivotal methodology that fine-tunes LLMs, especially instruction-optimized models. The RLHF method is based on the principles of RL, a subfield of ML in which an agent acquires decision-making skills through interactions with its environment, receiving either rewards or penalties as feedback. However, defining explicit rewards for good or bad responses in a conversation is intensive and can be impractical at scale. To improve how these models respond, we get help from people. AI trainers have conversations with the model. When the model gives different responses, the trainers decide which ones are best. They rank the answers from best to worst. This ranking gives the model a kind of score: good answers get high scores. Then, the model uses these scores to learn to give better answers, a process similar to honing your skills in a game by analyzing the scores you earn.

The following diagram outlines the process of RLHF:

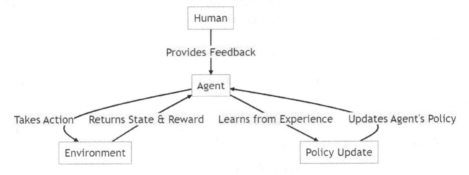

Figure 1.5 – The process of RLHF

This process is often iterative, with multiple rounds of interaction, feedback collection, and fine-tuning to improve the model's responses. Importantly, RLHF allows for more nuanced and context-sensitive feedback than traditional **supervised learning** (**SL**), leading to better alignment with human values and a decrease in harmful or inappropriate responses.

The challenges of RLHF

Despite its benefits, RLHF also comes with challenges and can produce harmful or factually incorrect text.

The process of gathering human preference data, an essential component of RLHF, can be costly due to the necessary inclusion of human workers. RLHF's effectiveness hinges on the quality of its human annotations, which include human-generated text and human-preference labels between model outputs. This data gathering often requires hiring additional staff, making it costly and challenging for academic labs. Furthermore, human annotators can disagree, adding a potential variance to the training data.

RLHF is a vital tool in creating effective, safe, and user-friendly instruction-optimized LLMs, but it requires careful implementation and ongoing refinement.

There is still considerable scope for further research to enhance RLHF's performance. There is also potential in open sourcing or crowdsourcing human assistance in training these models. At the time of writing, one successful example of open sourced RLHF is the Open Assistant model created by LAION, where contributors can assist with ranking, labeling, and making responses: `https://open-assistant.io/`.

The role of tokens in LMs

LLMs such as OpenAI's GPT-3 do not technically predict words; instead, they predict tokens. To understand this concept, it's important to recognize what tokens are and how tokenization works.

Tokenization is a common process in NLP that involves breaking down text into smaller units, called tokens. Depending on the tokenization strategy, these tokens can represent a whole word, a part of a word, or even punctuation.

In English, a token is often equivalent to a word, but this isn't always the case. Some tokenization approaches, such as WordPiece, **Byte-Pair Encoding** (**BPE**), or Unigram LMs, used by models such as GPT-3, break down words into smaller subword units. For instance, the word "conversational ai" might be split into five tokens: *["con", "versa", "ational"," a","i"]*.

When we say LLMs predict tokens, it means the model's task is to predict the most probable next token given a sequence of previous tokens. This token-level prediction allows the model to handle a vast range of vocabulary, including rare words, names, and even neologisms. It also helps the model to handle languages other than English more effectively, as many languages have complex morphology that can't be adequately represented at a simple word level.

This is crucial to LLMs' ability to generate coherent and contextually appropriate text. By predicting the next token in a sequence, the model can build up sentences and complete thoughts, giving the appearance of understanding and generating NL. However, it's essential to understand that these models do not truly comprehend the text they generate; they are probabilistic models that generate sequences based on patterns learned during their training.

Token costs serve as the foundation for determining the cost of OpenAI's LLMs, including ChatGPT, which have multiple models with varying capabilities and price options, which we'll cover in later chapters.

Understanding OpenAI's LMs

OpenAI has emerged as a pioneer in the domain of conversational AI with its ground-breaking LMs, GPT.

The evolution of OpenAI's LMs

OpenAI's GPT-3 serves as the foundational architecture for ChatGPT and has undergone a number of advancements since its release in 2020.

GPT-3

Unveiled in June 2020, GPT-3 set a new standard for language generation capabilities with its different base models: Ada, Babbage, Curie, and Davinci. Each variant had its unique characteristics.

GPT-3.5

GPT-3.5 was introduced as an optimized version for text-completion tasks, and one model was exclusively designed for code-completion tasks. The latest version in this series, `gpt-3.5-turbo`, was launched in March 2023. `gpt-3.5-turbo` is the most advanced GPT-3.5 model, specifically optimized for chat applications. Most importantly, it was a tenth of the cost of the `text-davinci-003` model. This model can handle up to 4,096 tokens, and its knowledge is up to date until September 2021. Also created is the `gpt-3.5-turbo-16k` model. This has the same capabilities as the standard `gpt-3.5-turbo` model, but it provides four times the context, allowing for longer and more complex conversations.

GPT-4

Introduced on March 14, 2023, GPT-4 has been dubbed as the most advanced of the OpenAI LMs. It boasts increased factual accuracy, multimodality with image processing capabilities, and creative outputs. It offers two model variants: `gpt-4-8k` and `gpt-4-32k`, distinguished by their context window sizes.

Key differences between GPT-3, GPT-3.5, and GPT-4

There are some key differences between the available models offering a range of capabilities, context capacity, and input types:

- **Capabilities**: GPT-4 significantly outperforms its predecessors in reliability, creativity, collaboration, and handling of nuanced instructions. Test comparisons conducted by OpenAI on different benchmarks have proven GPT-4's superiority in various domains.
- **Context length**: One of the most prominent distinctions between the GPT versions is the context length. While GPT-3 models could process a maximum of 2,049 tokens, GPT-3.5 improved this to 4,096 tokens. GPT-4 took a giant leap, with its two models capable of handling 8,192 and 32,768 tokens respectively.
- **Input types**: GPT-4 goes beyond processing text inputs and has the capacity to handle images. This unique ability can revolutionize its applications across domains.
- **Cost implications**: Greater capabilities come with a steeper price. GPT-4's advanced features command higher costs per 1K tokens for prompt and completion. The usage cost is not just higher but also more unpredictable due to the different costs for input and output tokens.

While GPT-4 undoubtedly sets new benchmarks in the realm of conversational AI, it's important to remember that it doesn't make GPT-3 and GPT-3.5 obsolete. Each of these models offers unique capabilities and potential applications.

Conclusion

In conclusion, the ChatGPT models are based on RLHF and the GPT series of models from OpenAI, which are themselves trained on extraordinarily large amounts of data.

To create ChatGPT, the latest GPT-3.5 Instruct model was fine-tuned with conversation examples instead of the whole internet to concentrate on improving the model's specialist conversation abilities. RL was then used so that the model could practice its conversational skills and improve its responses.

In the next sections, we'll explore in more detail its abilities, use cases, and transformative effect on the landscape of conversational AI.

Capabilities and applications of ChatGPT

LLMs offer a plethora of practical uses, with ChatGPT distinguishing itself through its advanced NLP capabilities. In this section, we unpack the array of applications that this technology brings to the table and a summary of the standout industry areas where ChatGPT holds the most potential. While many of these use cases are not necessarily new for conversational AI applications, it's true to say that ChatGPT can potentially take on more advanced tasks.

Capabilities of ChatGPT

In the following sections, we'll explore the various applications of ChatGPT, starting with NLU and **NL generation** (**NLG**) for chatbots and virtual assistants. Then, we'll move on to its capabilities in machine translation, summarization, sentiment analysis, and content creation. Additional topics include its use in spam filtering, language tutoring, and software development. Finally, we'll discuss the extended features available through ChatGPT plugins.

NLU and NLG for chatbots and virtual assistants

Conversational capabilities are the core focus of this book, and we will look at these in more detail.

The revolutionary capabilities of ChatGPT, particularly its ability to maintain complex multi-turn conversations across diverse topics and styles, mark a major milestone in the field of conversational AI. ChatGPT can be used to carry out what most conversational AI systems and experiences have often struggled to achieve. For seasoned practitioners who have been operating with intent-based systems and are aware of the challenges, the advent of ChatGPT has truly been transformative.

It must be noted that it's also exciting to see the coverage that conversational AI is now receiving across various media platforms, including TV, radio, and press.

As well as powering conversations, the less obvious use case is that ChatGPT can also be used as a tool to create and manage conversational AI applications, which are still powered by intent-based systems. We'll look in more detail at these applications in *Chapter 2*.

Machine translation between languages

ChatGPT supports 95 languages. The language translation technology in models such as ChatGPT bridges language barriers in sectors such as education and travel. By learning from past interactions, these systems deliver accurate real-time translations, enhancing cross-cultural communication.

Summarization of articles, reports, or other text documents

ChatGPT models excel at condensing complex topics into concise, easily understandable summaries. They can rapidly process large volumes of text, providing users with precise, digestible information and eliminating the need for manual research.

Sentiment analysis for market research or social media monitoring

By understanding and generating human-like responses, tools such as ChatGPT provide businesses with deeper insights into customer sentiments. They can be integrated into systems for analyzing and flagging potentially fraudulent activities, particularly in sectors such as banking.

Content generation for marketing, social media, or creative writing

Content creators can harness ChatGPT to generate compelling content efficiently. From blog posts and marketing materials to social media posts, the model generates unique, user-tailored content. It can even offer topic suggestions, proofreading, and editing services.

Spam filtering, topic categorization, or document organization

In any industry, ChatGPT can be used to categorize customer inquiries, detect suspicious transactions, and analyze content.

Personalized language learning and tutoring tools

The language support offered by models such as ChatGPT, which understands over 40 languages, extends to personalized language learning tools. They can handle inquiries and help in multiple languages, thereby enhancing accessibility and user experience in language learning.

Code generation and software development assistance

ChatGPT can understand and produce code in multiple languages, including the following:

- Python
- JavaScript
- C++
- C#
- Java

- Ruby
- PHP
- Go
- Swift
- TypeScript
- SQL
- Shell

ChatGPT simplifies debugging and code refactoring tasks, offering efficient solutions to complex coding problems. It can quickly locate potential problems, automate manual tasks, and provide highly accurate results, making it a valuable tool for software developers.

ChatGPT plugins have also extended these capabilities with the creation of their own code interpreter plugin aimed at handling some specific tasks:

- Solving mathematical problems
- Carrying out data analysis and visualization
- Converting files between formats

Working with images

ChatGPT can't generate images. However, when used with GPT-4, it can analyze infographics and images and answer questions based on the input. For example, you can input a picture of food ingredients and it will come up with a recipe, or it can describe an infographic.

When used with other technologies by using the new plugin capabilities, it's also possible to create graphics by passing an image description to these services.

ChatGPT plugins

ChatGPT plugins were released in March and are an innovative development by OpenAI, aimed at expanding the real-world applications, impact, and safety of GPT-4 and ChatGPT.

ChatGPT plugins serve as specialized extensions, crafted to augment the features of LMs. They provide ChatGPT with the ability to access real-time information, execute computational tasks, and interface with third-party services.

Developers can architect plugins for ChatGPT, and the model is furnished with specific instructions on leveraging each plugin through comprehensive documentation. Conceptually, plugins function as sensory inputs for LMs, empowering them to fetch information that is too recent, personal, or context-specific to be part of the original training data. They also permit the model to perform secure, bounded actions in response to users' explicit requests.

Plugins present a wealth of advantages. They address intrinsic challenges associated with LLMs, such as hallucinations (which we will cover later in this chapter), keeping abreast with recent developments, and accessing proprietary information sources with requisite permissions. By facilitating explicit access to external data, LMs can enrich their responses with reference-based evidence, enhancing the utility of the model and allowing users to cross-verify the accuracy of the output.

Plugins go beyond merely addressing current limitations; they pave the way for practically any use case. Collaborations between OpenAI and companies from diverse industries are revolutionizing our interactions with technology, from browsing product catalogs and making flight reservations to ordering products directly from organizations.

As the ChatGPT plugin ecosystem continues to grow, we can expect to see even more innovative applications and solutions that leverage the power of ChatGPT.

OpenAI released its own plugins, a web browser, and a code interpreter, as well as several third-party plugins from trusted providers.

How smart is ChatGPT?

AI, especially when it comes to LMs such as ChatGPT, isn't necessarily "smart" in the traditional human sense. So, basically, it doesn't have consciousness or self-awareness. Instead, it shows a high level of ability in understanding and generating human-like text, thereby appearing smart.

Recent studies have attempted to quantify the smartness of AI models such as ChatGPT, GPT-3, and GPT-4 by putting them through their paces with IQ tests and real-world applications.

OpenAI put GPT-3.5 and GPT-4 through several professional and academic benchmarks, as well as benchmarks designed for ML models.

The results are well documented here: `https://openai.com/research/gpt-4`.

The results show that these GPT models have proven proficiency levels at par with, and in some cases surpassing, the average human performance in various domains. For instance, ChatGPT achieved scores placing it in the 99.9th percentile on verbal-linguistic IQ and the Raven's ability test, which measures abstract reasoning and problem-solving skills.

The impressive achievements of ChatGPT don't stop there. It has successfully passed a Spanish medical examination and the **United States Medical Licensing Examination** (**USMLE**), outperformed college students on the Raven's Progressive Matrices aptitude test, and even passed an MBA degree exam at Wharton. In addition, GPT-3.5, a later variant of GPT-3, successfully passed the US bar exam and the US **Certified Public Accountant** (**CPA**) exam.

ChatGPT has also been used in real-world applications. It assisted a judge with a verdict in Colombia, wrote several bills in the US, and even made the cover of *TIME* magazine. The achievements of these AI models offer a remarkable testament to the potential of the technology.

The achievements of ChatGPT and its variants illustrate the profound capabilities of these AI models. However, while these scores and successes highlight their proficiency in understanding and generating text, it's crucial to remember that they don't equate to human-like intelligence or consciousness. AI's "smartness" lies in its programming and its ability to process and generate text based on that programming. The "smartness" of ChatGPT, as demonstrated by its high IQ scores and real-world applications, is a testament to the impressive strides made in the field of conversational AI.

Applications of ChatGPT

Moving on from the applications of traditional conversational AI approaches and, specifically, more intent-based approaches, let's look at some real-world applications of ChatGPT.

Business and finance

The adoption of AI in the business and finance sector has seen a significant rise in the recent past. From automating financial report generation to identifying potential financial risks, ChatGPT has proven to be a valuable tool in streamlining operations and contributing to data-driven decision-making:

- **Customer service chatbots**: A critical application of ChatGPT, conversational AI can be leveraged in any role where customer support needs to be handled intelligently and at scale. ChatGPT provides a front-line interaction platform, creating agents equipped to handle customer inquiries, process transactions, and even offer personalized product recommendations.

- **Market analysis and forecasting**: ChatGPT's capabilities extend to the domain of market analysis and forecasting, where it can analyze vast volumes of financial data to identify patterns and trends and provide insights into market conditions. It can generate analytical summaries that offer a comprehensive understanding of the financial landscape.

- **Investment management**: ChatGPT's data analysis capabilities also make it a valuable tool for investment management. It can help businesses and investors make informed decisions by offering personalized investment recommendations that consider individual risk profiles and financial goals.

- **Fraud detection**: In the realm of financial security, ChatGPT can be leveraged to detect fraudulent activity. By analyzing transaction data and identifying patterns indicative of fraudulent activities, ChatGPT can provide robust fraud detection mechanisms.

- **Risk management**: ChatGPT is also instrumental in risk management. By analyzing financial data and identifying potential risks, it can help businesses and financial institutions devise strategies to mitigate those risks.

- **Financial reporting**: By analyzing data and providing insights into financial performance, ChatGPT can be used to automate the generation of financial reports.

To sum up, the applications of ChatGPT in the business and finance sector are vast and transformative. From augmenting customer service with chatbots to identifying risks and fraud, this AI model offers a myriad capabilities, revolutionizing traditional business operations and paving the way for a future dominated by smart, data-driven strategies.

Healthcare and medical applications

Probably an area fraught with the most ethical and regulatory challenges. In healthcare, ChatGPT's robust conversational experiences can be employed in healthcare and medical applications. From helping in diagnoses and treatment planning to driving patient engagement, there are many use cases in this area:

- **Chatbot for patient triage**: ChatGPT can be an instrumental tool in the development of chatbots designed for patient triage. Such systems help healthcare providers assess the urgency of a patient's condition and find the most appropriate course of action.

- **Medical diagnoses and treatment recommendations**: One of the compelling applications of ChatGPT lies in aiding with medical diagnoses and treatment recommendations to help improve the accuracy and efficacy of treatment plans. By analyzing patient data, symptoms, and medical histories, ChatGPT can provide medical professionals with valuable insights and suggestions for diagnosis and treatment.

- **Medical education**: ChatGPT's ability to supply accurate and comprehensive information makes it a valuable tool for medical education. It can serve as a resource for healthcare providers and patients alike, offering information on a wide array of medical conditions and treatment options.

Employing ChatGPT in the role of a trainer or interviewer is also where the technology can shine:

- **Mental health counseling**: ChatGPT can serve as a base for chatbots designed to provide mental health counseling. By analyzing patient data and offering personalized recommendations, these chatbots can aid patients in managing their mental health conditions. Although not a replacement for professional counseling, these AI-enabled platforms can give initial support and guidance.

  We've seen conversational AI technology successfully implemented in this use case in a number of chatbot products, such as Woebot (`https://woebothealth.com`) and Wysa (`https://www.wysa.com`).

- **Patient engagement and adherence**: ChatGPT can also be utilized to bolster patient engagement and adherence to treatment plans. It can generate personalized reminders and recommendations, aiding patients in adhering to their prescribed regimen. This form of AI-driven patient engagement is particularly beneficial for chronic disease management where medication compliance is key.

The applications of ChatGPT in healthcare and medicine are multifaceted and impactful. By providing reliable and data-driven solutions in diagnosis, patient engagement, and research, ChatGPT is paving the way for a new era of AI-assisted healthcare.

Law and legal services

AI is increasingly penetrating all sectors, and the legal domain is no exception. ChatGPT's applications range from summarizing legal documents and assisting in drafting legal briefs to providing insights for legal research and facilitating efficient communication among legal professionals:

- **Contract review**: ChatGPT can play a crucial role in scrutinizing contracts to pinpoint legal concerns such as inconsistencies or vague terms that need further examination or amendment.

- **Legal advice chatbots**: Again, another conversational AI-specific use case. ChatGPT can form the basis of legal advice chatbots capable of helping clients with legal inquiries. A ChatGPT-powered agent can analyze legal data and provide personalized recommendations to help clients understand their options.

- **Document drafting**: ChatGPT can be used to streamline the drafting of complex legal documents such as contracts and briefs by analyzing relevant legal data and proposing well-informed content and suggestions.

- **Due diligence and e-discovery**: ChatGPT can be a powerful tool for conducting due diligence and e-discovery, capable of assessing legal documents, executing background verifications, and recognizing probable legal challenges, as well as helping to pinpoint pertinent documents and data during legal disputes by scrutinizing textual data and identifying prevailing trends and patterns.

In conclusion, the implications of ChatGPT for law and legal services are profound. It offers transformative solutions that augment the capabilities of legal professionals, enabling them to perform their duties more efficiently and effectively. From facilitating legal research and contract reviews to streamlining the drafting of legal documents and expediting e-discovery, ChatGPT is indeed revolutionizing the field of legal services, shaping a future where AI and legal practice are intrinsically intertwined.

Education and training

Education and training are a hot topic when it comes to ChatGPT and LLMs in general. As the LLMs have gotten smarter, so have their applications in education and training, from providing real-time feedback, generating engaging educational content, assisting educators, and creating adaptive learning environments:

- **Personalized learning**: ChatGPT can play a critical role in personalizing learning experiences. By using data on learners' preferences, strengths, weaknesses, or ability levels, ChatGPT can generate recommendations for learning materials and activities, adhering to a personalized program in any specific domain.

- **Teacher support**: Similarly, ChatGPT can assist educators significantly by suggesting targeted recommendations for lesson plans based on specific age, curriculum, or level and providing different techniques for classroom management and teaching approaches.

- **Test preparation**: ChatGPT can be an essential tool for test preparation. It can offer recommendations for study materials, as well as create practice exams and test papers.

These capabilities can also enable rich learning experiences when coupled with new mediums such as digital humans.

In conclusion, the intersection of AI and education opens a multitude of possibilities. The use of AI LMs such as ChatGPT in the educational sector is having profound effects on teaching and learning practices. From personalized learning and teacher support to language learning, test preparation, and online tutoring, ChatGPT's applications are helping to reshape the educational landscape, making it more personalized, engaging, and effective. As we move forward, we can expect to see even more innovative uses of AI in education, ushering in a new era of learning empowered by technology.

Many of the applications listed here are like the use cases covered in an earlier part of this chapter, *Understanding conversational AI applications*. It's no surprise that conversational AI solutions have been developed for many of them in the past with varying degrees of success.

Limitations of ChatGPT

ChatGPT, as with other LLMs, is not fallible and has several significant risks, limitations, and security considerations that are important to be aware of if you are going to use ChatGPT in production conversational AI use cases.

Limitations

There is no doubt that ChatGPT has some incredible capabilities, but let's take a look at where things can go wrong.

Accuracy and depth

Probably the most well-known limitation and the one that has garnered the most effort to overcome is ChatGPT's *"information cutoff"* date. The early GPT-4 models were trained on publicly available internet data, with cutoff dates as far back as 2021. The latest GPT models have shown improvements. According to OpenAI, the newest GPT-4 Turbo (gpt-4-turbo) has a knowledge cutoff of December 2023. However, beyond this point, it is not updated with the latest information or events. This limitation means that conversational AI implementations using ChatGPT cannot natively support data or events occurring after December 2023, at least until a new model is released.

The ability for ChatGPT to answer questions about private data or more recent information is, of course, a core requirement for many conversational AI applications, which we'll look at achieving in more detail later.

Hallucinations and accuracy

In the context of LLMs, *"hallucinations"* refer to when the model generates text that is not coherent, relevant, or accurate. The problem of hallucinations in AI systems such as ChatGPT is a complex one. This can happen because ChatGPT has learned certain patterns in the data that do not apply to the specific context of the task. In other cases, an LLM may generate text that is completely unrelated to the input or task, which can be described as *"hallucinating"* text. For ChatGPT, hallucinations are one of the most concerning issues.

It's not just about the AI making things up; it's about the AI generating information that could potentially be harmful or misleading. This is particularly problematic in scenarios where the AI's output is used to make important decisions or to inform people about critical issues.

One of the main concerns is the accuracy of outputs generated by ChatGPT. OpenAI has acknowledged that the outputs can sometimes be inaccurate, untruthful, or misleading. This is a complicated issue for companies aiming to rely on production applications powered by ChatGPT.

Accuracy is not just a consideration in ChatGPT conversational AI applications. Anywhere generated content comes into scrutiny and needs to be factual can be problematic. In June 2023, two US lawyers were fined for submitting fake court citations from ChatGPT. The lawyers' firm was penalized after it transpired that fake quotes and citations were created by ChatGPT and submitted as part of an aviation injury claim.

While ChatGPT is adept at generating text that appears coherent and logical, its responses may not always be correct or detailed enough. Output often needs manual review and robust validation mechanisms. It's crucial to address the issue of hallucinations and be constantly aware that ChatGPT cannot discern factual accuracy to ensure the reliability and safety of ChatGPT applications.

Context limitations

ChatGPT manages conversation context by using earlier conversation information for each new turn of the conversation. This way, it knows what's been said previously, and it can use this as part of its process of deciding what to say next.

Despite improvements in context understanding, both GPT-3.5 and GPT-4 have context limits, with GPT-4 having a significantly larger limit. This limitation can lead to instances where ChatGPT only considers its most recent output and overlooks context from earlier in the interaction. OpenAI has also recently released a new chat-specific model: `gpt-3.5-turbo-16k`, which has a context four times larger than the earlier one. Context limitations are an area that can be improved with a summarization strategy and the approaches covered later in the book.

Risks and security

The rapid adoption and widespread use of ChatGPT have raised significant security and **intellectual property (IP)** concerns that the technology could be misused by criminal enterprises.

Property rights

Another complication is a company's IP rights, especially when claiming rights to something generated by ChatGPT or partially AI-generated.

According to OpenAI's terms of use, the user owns the input, and OpenAI assigns all its rights, title, and interest in and to the output. However, OpenAI cannot assign rights to content it did not initially own. This has led to unresolved questions about AI-generated IP and its ownership.

Security

Not necessarily a security issue related to AI and ChatGPT functionality per se, but a very real one considering the number of users OpenAI has. ChatGPT's own security was put to the test in 2023 when OpenAI confirmed a data breach caused by a vulnerability in the code's open-source library. This breach allowed users to see the chat history of other active users, raising concerns about the security of user data. The breach was quickly patched, but it highlighted the potential risks associated with the use of AI technologies.

In another incident, a deeper investigation revealed that the same vulnerability had likely exposed payment information for a few hours before ChatGPT was taken offline. This incident raised further concerns about the security of sensitive user data.

The rising popularity of ChatGPT plugins could also amplify safety hazards by inadvertently enabling harmful or unintended actions despite OpenAI's best efforts.

Educational concerns

As covered earlier, ChatGPT can pass educational tests and carry out real-world applications. This has resulted in concerns among education experts and academic faculties, due to ethical issues, cheating, and misuse in schools.

Many academic institutions have banned the technology on their networks, which is a futile measure but shows that there is real concern that students might exploit ChatGPT to write papers, cheat, and sidestep the learning process. This raises questions about the integrity of academic institutions and the very nature of learning and education. As the capabilities of LLMs grow, these issues are only set to increase despite the growth of other LLM technologies to spot AI-generated content.

Privacy concerns

The use of AI LMs such as ChatGPT also raises privacy concerns. These models store vast amounts of data and use this information to generate responses.

Content already used to train the model is a major concern, but one more pressing is what happens with ChatGPT conversation data.

If sensitive data, such as an organization's IP or sensitive customer information, is entered into the chatbot, it enters the chatbot library and is out of the user's control.

In response to these concerns, some businesses and countries have tightened restrictions on AI use. For instance, JPMorgan Chase and Apple have restricted employees' use of ChatGPT over concerns about security.

Italy has temporarily blocked the application across the country due to **General Data Protection Regulation (GDPR)** compliance concerns.

It's worth noting that ChatGPT allows users to stop chat history logging, but it's an important consideration if users are using internal data such as medical records as part of their prompt engineering.

Malicious usage

Many of the strengths of ChatGPT, such as realistic conversational ability and text generation, are likely to be used for criminal intentions, from enhanced phishing on an industrial scale to targeted disinformation.

Criminals will also be empowered to use less traditional mediums such as messaging platforms to carry out targeted attacks.

In the past, phishing emails were often characterized by poor grammar and unusual sentence structure, making them easy to identify. Criminals will leverage ChatGPT to develop convincing phishing emails. These emails can now mimic native speakers and deliver tailored messages. ChatGPT's ability to seamlessly translate languages is expected to significantly aid foreign criminals' abilities to carry out global attacks.

Bias

ChatGPT, as with many other LLMs, has been trained on human-generated content from the internet. As a result of this, ChatGPT is susceptible to various biases, including gender, racial, cultural, language, ideological, and commercial, as well as several others.

Bias issues are wide-ranging; here are a few examples:

- Generating content that fails to incorporate a well-rounded perspective on diverse personal experiences and linguistic nuances
- Crafting responses that allocate roles or jobs based on gender or ethnicity, thereby fortifying pre-existing stereotypes
- Creating text that portrays a subtle bias toward certain social philosophies or political ideologies
- Producing content that mirrors the dominant narratives present in the training data, resulting in the creation of sensationalist or attention-seeking titles and statements

Even with safeguards in place, these models will sometimes say sexist/racist/homophobic things. Be careful when using LLMs in consumer-facing applications, and care needs to be taken when using ChatGPT in research.

Summary

In conclusion, in this chapter, we've seen how conversational AI has made steady progress from rule-based to AI-driven intent-based systems. By looking at the applications of ChatGPT, we've seen that as the capabilities of modern conversational AI systems have grown, so have the breadth and depth of use cases for conversational experiences. There are thousands of these systems, carrying out complex tasks across multiple industries.

From our initial focus on ChatGPT, we have seen it's a powerful tool that has the capabilities to revolutionize the field of conversational AI by taking on more complex tasks and use cases that until recently have been difficult to achieve with modern conversational AI platforms. It offers a wide range of applications in many industries, including customer service, healthcare, finance, law, education, and more.

However, the chapter has also highlighted that it's important to be aware of its limitations and risks, including accuracy and depth limitations, hallucinations, context limitations, security concerns, privacy issues, and biases. These challenges require careful consideration and mitigation strategies when developing and deploying ChatGPT applications. As we move forward, it is crucial to address these limitations and ensure that ChatGPT is used responsibly, ethically, and with proper safeguards in place.

Before we start looking at creating conversational applications with ChatGPT, let's look at how the technology can be used in traditional conversation design.

In the next chapter, we will delve deeper into the conversation design applications of ChatGPT. We will explore how to design effective and engaging conversations using ChatGPT, covering topics such as dialogue flow, user prompts, system responses, context management, and more. Join us as we uncover the practical skills needed to create powerful and meaningful conversational AI experiences using ChatGPT.

Further reading

To explore more on the topics related to this chapter, you can visit the following links:

- Enterprise conversational AI

  `https://www.gartner.com/reviews/market/enterprise-conversational-ai-platforms`

- *Bridging the Gap: A Survey on Integrating (Human) Feedback for Natural Language Generation*

  `https://arxiv.org/abs/2305.00955`

2

Using ChatGPT with Conversation Design

In this chapter, we will delve into the intersection of ChatGPT and conversation design. We'll look at the role of the conversation designer and explore how **large language models** (**LLMs**) such as ChatGPT have impacted conversation design. We will also examine the practical applications of ChatGPT in conversation design, including its use in simulating conversations and creating personas. Finally, we'll discuss the importance of testing and iterative design in creating effective and engaging conversational AI systems. By the end of this chapter, you will have a comprehensive understanding of how to utilize ChatGPT in conversation design.

Some of the prompt results are listed, while others have been omitted in the interests of space. We encourage you to try the prompts for yourself.

In this chapter, we're going to cover the following main topics:

- Understanding conversation design
- Working with practical applications of ChatGPT in conversation design
- Simulating conversations
- Persona creation with ChatGPT
- Testing and iteration in conversation design with ChatGPT

Technical requirements

In this chapter, we will be using ChatGPT. We recommend that you sign up with OpenAI.

Understanding conversation design

Conversation design is a complex and involved role that is key to creating a chatbot or voice assistant that is helpful and natural and, as a result, successful. There are other roles in the field of conversation design, particularly that of an AI trainer and variations of this role. For the purpose of this book, we will consider these other roles when discussing conversation design, as there is considerable overlap between the tasks carried out for many of them.

Conversation design leverages the principles of user experience design, linguistics, cognitive psychology, and AI to create engaging, interactive experiences between humans and machines over the medium of voice or text. At their core, conversation designers understand communications, the technology landscape of conversational AI, the mediums to deploy to their users, and how to satisfy user requirements.

The goal of conversation design is to create a smooth, intuitive, and conversational user interface that feels natural to the user with the technologies available to mimic human-like conversation.

Exploring the role of conversation designers

Conversation designers carry out several important tasks. These are important to cover so we can see where LLMs and ChatGPT can be used to streamline and improve the process. Let's look at the processes involved in creating conversation designs. It's these designs that will be implemented on the chosen conversational AI platform for the project.

Understanding user needs

The first step in conversation design is to understand the user's needs and behaviors, as well as the context in which the conversation takes place.

A conversation designer's role is to understand an organization's operations, processes, challenges, and aims. A good designer can turn business requirements into customer journeys, use cases, and natural flows. The designer needs to understand the personas of the intended users, including their age, demographic, needs, expectations, and language. They also need to understand the current interactions within the organization or application. A thorough understanding of existing inquiries and existing unstructured or structured data is especially important here, as is a solid understanding of the channel the conversational experience will be deployed to.

Designing the persona

The persona for the agent not only decides its tone of voice but also guides its approach to conversations. This needs to be carefully considered to match user needs.

Existing brand and content guidelines will also need to be considered.

For instance, a chatbot deployed on an enterprise software platform to handle customer support queries is going to have a different tone of voice and persona than an agent designed to help students with queries on WhatsApp during orientation week .

Designing conversations

The conversation flow is the sequence of interactions between the user and the conversational agent.

A conversation designer will need to prioritize existing use cases to understand what jobs an agent needs to carry out and how a conversation can be structured to achieve this task.

Intents clarification, utterance creation, and entities clarification

An intent represents the user's intention. For example, the intent could be to find out the status of an order or to book a flight. Each intent is associated with a set of utterances, which are the phrases that users might use to express that intent. For example, for the intent of checking flight status, the utterances could be "*When does my flight leave?*" or "*Is my flight on time?*".

Entities are the specific pieces of information that the chatbot needs to fulfill an intent. For example, for the intent of booking a flight, the entities could be the departure city, arrival city, and date of travel. It's the conversation designers' job to understand each type of entity and the different entries for each one. Common entities are often straightforward, as they are supported by conversational AI models. However, if a custom entity is needed, each one will have to be sourced. For example, a dog-training chatbot may need to understand all the different dog breeds.

Dialogue design

With the user journeys mapped out for the conversation, it's the conversation designers' job to create the details of the interactions. This involves designing the conversation flow between the user and the conversational agent. It includes defining the prompts that the chatbot will use to guide the conversation, the user responses, and the agent's replies. The dialogue should be designed to be natural and engaging, and it should guide the user toward achieving their goal.

Designers carefully script dialogue and responses with different paths. These are based on potential user inputs where information needs to be gathered to enable the user to convey their needs, and the agent to gather the values needed to answer the query. Designers start with what is commonly known as the happy path, which is the easiest route through conversation to achieve the goal, and then build out to handle the edge cases around a conversation.

The design needs to support other situational edge cases, such as unhandled intents, mismatched intents, on-input, and any other ways a user can try to steer a conversation. There is a lot to document here. These designs can become complex with many turns of a conversation, as conversations are often not linear, and users may change topics or ask unexpected questions. It's the conversation designer's job to include this in a design.

Conversation designs are shared between project stakeholders using several different methods, as follows:

- Flow charts
- Spreadsheets
- Pseudocode
- Conversation management software, such as Voiceflow

Depending on the size of the project, team, and budget, as well as personal preference, there is no one-size-fits-all approach for conversation design tools. As long as they are easy to iterate and collaborate on, these designs will often be touched by several different teams on the way to sign-off. However, some of the more popular conversation management tools offer a lot more than just design and collaboration and include some very useful features leveraging LLMs. This will be covered in the next section.

Copywriting

The last step is to write the actual text of the chatbot's responses. This involves using persuasive and engaging language to guide the user through the conversation. The copy should reflect the chatbot's persona and align with the brand's voice and tone. For larger agents with hundreds or even thousands of responses, the challenge is often in securing consistency, particularly if there are several conversation designers working on a project in tandem or over time. As a conversational AI project matures, it's even more likely that this will occur. As each party brings their own style to the project, the tone of voice can deviate from the original persona and brand guidelines for the agent.

Now we've covered the role of a conversation designer and the specific tasks that make up the process of conversation design before implementing a conversational AI solution, it's time to see where ChatGPT can be utilized in these specific workflows.

Working with practical applications of ChatGPT in conversation design

From our understanding of the tasks that need to be carried out for conversation design, it's easy to see that in some areas, we can leverage ChatGPT to help complete them. Some of the most popular conversation design tools also supply the ability to carry out these tasks as part of their functionality.

> Tip
> Remember that ChatGPT is a conversation design tool, not a replacement for conversation design best practices.

Intent clustering with ChatGPT

Intents are the core concept of most traditional conversational AI agents. The intent is there to define the intention of the user, and each intent needs to have a set of utterances to fire this intent. Many conversational AI implementations start with large volumes of live chat data, voice transcripts, or other unstructured text data, which will form the basis of the tasks or intents that a chatbot or voice interface will need to support.

How can intent clustering be useful?

A conversational AI project may be in its infancy. It may also be a mature project under heavy use or one whose subject matter changes often. It's often the conversation designer or AI trainer's job to use this data to understand and prioritize the intents by clustering unstructured training data or utterances.

For example, if you are looking to move from live chat to chatbot automation, you may have a large corpus of historical chat. You'll want to look at this and decide which common questions or intents are asked the most. Similarly, a chatbot or voice assistant that handles a specific fluid subject matter, such as holidays, may need to be changed regularly to answer new questions arising from external factors. These new questions could be highlighted by clustering.

What is intent clustering?

Intent clustering essentially means looking at a large corpus of data and looking for groupings of semantically similar sentences. So, for our use case, we will be looking at our unstructured chat data and grouping based on semantically similar meanings so that we can start surfacing intents.

In the past, the process of intent surfacing by clustering has involved either a complex process using embeddings, or the use of proprietary solutions such as HumanFirst. The process would follow these steps:

1. **Data collection**: Assemble the corpus of text data that you want to analyze. This could range from a collection of scientific articles to a dataset of user-generated content.

2. **Embedding generation**: Use machine learning libraries such as TensorFlow, services such as Word2Vec, or **Bidirectional Encoder Representations from Transformers(BERT)** to generate similarity embeddings for your text data. These embeddings are high-dimensional vector representations that capture the semantic content of the text. So, if a text has a similar meaning to another, then the embeddings will be closer together

3. **Clustering algorithm**: Apply an unsupervised machine learning algorithm such as K-means to the generated embeddings. These algorithms separate the high-dimensional embedding space into clusters based on the density and distance between points.

4. **Cluster analysis**: Once complete, there should be sets of clusters of utterances, each containing semantically similar texts. Analyzing these clusters can reveal intents or questions, which can then be used to train an intent-based machine learning model.

In essence, clustering in this context is an unsupervised technique that uses semantic embeddings to group similar texts together, helping the discovery of underlying intents in large text datasets of unstructured text data.

Learning how to create intent clusters with ChatGPT

Looking at achieving this with ChatGPT is straightforward when you follow the simple rules of prompt engineering: instruction, context, input data, and output indicator. We'll cover more examples of prompt engineering in more detail in *Chapter 4*.

Prompting, for our intents, could look something like the following. It's important to note that designing prompts is an iterative process that sometimes requires experimentation to get the results you are looking for.

For our first example, we are looking at creating a chatbot for our college website that will answer questions about the college. We have a set of unstructured utterances that have been harvested from a live chat implementation.

For this example, we are using instructions, input data, and an output indicator in our prompt. The instructions and input data handle telling ChatGPT what we want and what utterances to use. The output indicator also shows the format we want the result to use. In this case, JSON will be the most useful, as we're looking at using the intents or utterances as part of an automated process. The model would also handle other output formats such as CSV. It looks something like this:

```
Look at the following utterances from livechat transcripts and cluster
them into intents
Utterances:    "can you tell me where is the college located",
"where's the college", "where is college located", "how can I reach
college","whats is the college address", "whats the Syllabus for
IT", "what is the Information Technology syllabus"," "Whats my next
lecture"
Output: Return as JSON in the following format-{"intent":
{name},"utterance":{utterances}}
```

The result should look something like the following:

```
[
  {
    "intent": "Get_College_Location",
    "utterances": ["can you tell me where is the college located",
"where's the college", "where is college located", "how can I reach
college"]
  }, {
    "intent": "Get_College_Address",
    "utterances": ["whats is the college address"]
  }
]
```

Along with manually creating intents via the ChatGPT interface, it would also be possible to automate this process further by using the ChatGPT API directly.

Expanding the example to create intents

Considering our chatbot use case, it may be that we want to build our intent set by creating more intents. For example, if your live chat data is limited and you would like to expand out to other intents, you can use ChatGPT to carry out a task to create more intents and utterances using some of your existing ones.

Let's create our prompt by using the output from our earlier example, prompting for intent clusters, which you can see in the prompt with the `{input_data}` placeholder:

```
Look at the following intents and utterances for a chatbot which will
answer questions about a college, these JSON objects consist of an
array of intent objects, each of which includes a name and a list of
corresponding utterances.
Create another 5 examples of intents and ensure that each intent has
10 utterances:
Intents:
{input_data}
Output: Return as JSON in the following format-{"intent":
{name},"utterance":{utterances}}
```

If we run this prompt, we can see that this works well. This is the output:

```
[
    {
      "intent": "Get_College_Events",
      "utterances": ["Whats happening on Tuesday","What are the upcoming
events?", "Can you tell me about any college events?", "What's
happening on campus?", "What events are scheduled this month?", "Is
there a college event this week?", "Are there any sports events coming
up?", "When is the next cultural fest?", "Are there any seminars or
workshops?", "What's the college event calendar?", "Are there any
academic events happening soon?"]
    },
    {
      "intent": "Inquire_About_College_Facilities",
      "utterances": ["What facilities are available in the college?",
"Does the college have a library?", "Can you tell me about the sports
facilities?", "What are the lab facilities in the college?", "Are
there any on-campus cafeterias?", "Does the college have a gym?",
"Are there any hostel facilities?", "Does the college have a swimming
pool?", "Tell me about the medical facilities", "Are there any
recreational facilities in the college?"]
    },
    ...
    ]
```

Understanding utterance and entity generation

In modern conversational AI systems, intents and entities are crucial components that enable the system to understand and process human language, enabling more natural and effective conversations. The task of creating these elements is one of the most common tasks for a conversation designer. These elements can then be used to train our machine learning model for the conversational AI platform we are using.

Creating utterances

In the earlier *Learning how to create intent clusters with ChatGPT* section, we looked at clustering utterances to surface intents from unstructured training data. ChatGPT can also be used to create utterances.

If we don't have enough utterances for a specific intent, we can prompt ChatGPT to create a set of utterances for a specific intent, as follows:

```
Consider a "Get_College_Events" intent for a chatbot which will answer
questions about college events.
Create another 10 examples of utterances for each intent:
Output: Return as JSON in the following format-{"intent":
{name},"utterance":{utterances}}
```

The results are surprisingly good:

```
[
  {
    "intent": "Get_College_Events",
    "utterances": ["What's happening in college this week?", "Tell me
about the upcoming college events.", "Are there any events in college
next month?", "Is there a sports meet this year?", "What are the
details of the cultural fest?", "When is the next science fair?", "Is
there any guest lecture scheduled soon?", "When is the annual college
fest?", "Are there any workshops I can attend?", "Can you give me a
list of events for this semester?"]
  }
]
```

We can also be more specific about the types of utterances. For example, if we wanted a larger spread of utterances, we could prompt ChatGPT for longer tail examples. Let's use the "Get College Location" intent use case and add long tail utterances to the existing utterances:

```
Look at the following intent for a chatbot that will answer questions
about a college, these JSON objects consist of an array of intent
objects, each of which includes a name and a list of corresponding
utterances.
Create another 5 examples of long tail utterances for each intent and
add to the array of utterances
Utterances:
[
  {
    "intent": "Get_College_Location",
    "utterances": ["can you tell me where is the college located",
"where's the college", "where is college located", "how can I reach
college"]
  }
]
```

```
Output: Return as JSON in the following format-{"intent":
{name},"utterance":{utterances}}
```

This results in a larger breadth of utterances for our NLU training dataset:

```
[
  {
    "intent": "Get_College_Location",
    "utterances": ["can you tell me where is the college located",
"where's the college", "where is college located", "how can I reach
college", "I am new to this city, can you guide me to the college?",
"Can you provide the exact location of the college?", "I am not
familiar with the area, could you help me find the college?", "I would
like to visit the college, can you tell me its location?", "I need
directions to the college, can you help?", "Could you give me the
location details of the college?"]
  }
]
```

Each of the created utterances matches our criteria with a more conversational and complex (long tail) format.

> **Tip**
> Remember that artificial training data, even that created by LLMs, is no replacement for real usage data. This is particularly true when we are creating utterances for voice, as ChatGPT has been trained on textual data.

Creating entities

Another generative task that ChatGPT is very capable of is creating entities.

Entities are the specific pieces of information that relate to the intent. They are the variables that supply the details of the user's request.

For example, consider our Get_College_Events intent. To answer the question correctly, we need to know a couple of things: the dates and types of events. While system dates tend to be supported by the conversational AI platform, custom entities often need to be created to train the NLU. For this particular use case, the event type would be the custom entity we need to populate. We can ask ChatGPT to help us with this. Remember to be as specific as possible with the prompt. In this case, we're asking for the output to be formatted correctly for importing into our conversational AI platform:

```
Consider a "Event_Type" entity which needs to be extracted from
utterances for the Get_College_Events for a chatbot which will answer
questions about college events.
Create 5 examples of entity_val and 5 synonyms  for each entity type:
```

```
Output: Return as JSON in the following format-{"entity_val":
{entity_val},"synonyms  ": [synonyms]  }
```

The results are a comprehensive set of entity types and corresponding synonyms:

```
[
  {
    "entity_val": "seminar",
    "synonyms": ["workshop", "conference", "symposium", "meeting",
"forum"]
  },
  {
    "entity_val": "sports meet",
    "synonyms": ["athletic meet", "sports event", "sports day",
"athletic event", "games"]
  },…
]
```

By correctly finding and processing entities, ChatGPT can create precise and useful responses quickly.

In the absence of any customer conversational data, ChatGPT utterances can serve as a good means to bootstrap a chatbot for further iterations.

Using ChatGPT to help write your dialogue

Writing engaging chat and voice dialogue is one of the critical parts of a conversation designer's role. So, how can ChatGPT help, and should you even consider using it?

Understanding the challenges of writing dialogue

As conversation designers, one of the most engaging yet challenging aspects of our work is crafting the responses of our chatbot or voice assistant. After all, this is your chance to shine. Good dialogue is the difference between a flat conversational experience and one that is truly engaging for your users.

Creating sample dialogue, ensuring smooth transitions between bits of dialogue within a flow, and formulating precise responses for highly specific contexts all take time and careful consideration. This is all done while dealing with the pressure of meeting deadlines and keeping stakeholders informed and feedback actioned. Some of the things to consider include the following:

- Upholding a voice and tone that resonate with your AI assistant's persona
- Keeping to company content guidelines
- Utilizing language that aligns with the content you may need to offer your users
- Handling error scenarios capably and consistently
- Striking a balance between saying too little and being overly verbose

- Ensuring that the crafted responses are effective across various modalities, not just in theory

- Addressing inherent biases that could influence the language we use and impact the end users

- Following legal and compliance guidelines carefully to ensure the language we use is correct and not going to land you legal hot water

In essence, editing conversational dialogue is a complex set of tasks that requires a delicate balance of creativity, precision, and compliance.

Editing your dialogue with ChatGPT

ChatGPT can perform several tasks to enhance your dialogue. Let's delve into a few examples, such as rewriting for readability, checking content guidelines, or creating different versions of copy for a better conversational experience.

Rewriting for readability

If you have copy that is not written for your use case, but you still need to convey the included information, you can ask ChatGPT to rewrite it for you. Here is an example of a prompt for doing this. This is a copy of a patient information sheet for a clinical trial:

```
I'm writing a chatbot response for my "what is the purpose of
this drug trial" intent. Revise this dialogue to make it more
understandable for clinical trial patients and ensure it's under 100
words:
Chronic kidney disease (CKD), affecting 1 in 10 UK adults, is a
progressive loss of kidney function that can lead to complete kidney
failure, requiring dialysis or transplantation. High blood pressure
treatment, especially with drugs like Angiotensin Converting Enzyme
inhibitors (ACEi) or Angiotensin Receptor Blockers (ARBs), can slow
CKD progression. However, recent research suggests that stopping these
drugs in advanced CKD (stage 4 or 5) can stabilize kidney function
and delay dialysis, indicating these drugs might contribute to kidney
function decline. To confirm this, the ACEi trial will recruit 500
CKD stage 4 or 5 patients on ACEi/ARBs treatment, randomly allocating
half to continue and half to stop their treatment. The trial will also
examine other effects of stopping these drugs, such as heart effects
and participant quality of life.
```

From the result, you can see that ChatGPT has done an excellent job of summarizing the content:

```
The STOP-ACEi trial is for CKD patients, a kidney condition affecting
1 in 10 UK adults. We're studying if stopping certain blood pressure
drugs (ACEi/ARBs) can help stabilize kidney function and delay
the need for dialysis. We'll have two groups: one continues the
medication, the other stops. We aim to understand the best approach
for CKD treatment
```

Try this with your own example text. Keep iterating on your prompt until you have something you're satisfied with.

Another good dialog creation example is rewriting content for a specific purpose. For example, if we want to create a set of concise instructions from descriptive copy, we might use the following:

```
I'm writing copy for my dog training chatbot intent "how do i teach
my dog to sit". Rewrite these instructions into a set of steps in the
style of Barbara Woodhouse. Each step can only be 150 characters. Also
write an introductory sentence:
Training a puppy to sit is one of the most basic commands and is
useful in various situations. Start by holding a treat close to
your puppy's nose and then move your hand up, allowing their head to
follow the treat and causing their bottom to lower. Once they're in
sitting position, say "sit", give them the treat and share affection.
Remember, repetition is key. Practice this command a few times each
day and be sure to reward your puppy for sitting even without command,
to reinforce this behavior.
```

Providing concise instructions with context, input, and output instructions yields good results and something we can work with. Try this with your own examples.

It's tempting to think that we can automate our response creation. However, it's not wise to do so. The content created by ChatGPT is only a starting point. It's essential to ensure that ChatGPT is supplying the right information and not changing meaning while also keeping to your conversation design requirements.

Persona creation with ChatGPT

User personas are critical components in user-centric design, as they enable organizations to understand their target demographics and construct solutions tailored to their needs. This section will delve into how ChatGPT can be used to generate sophisticated and intelligent user personas.

> **Real-world example**
>
> Meet LuxeLocks, our imaginary company that specializes in selling luxury haircare products. LuxeLocks is passionate about providing its customers with high-end, quality products that transform their hair into a crowning glory. In an effort to elevate their customer service to the next level and to answer queries around the clock, LuxeLocks has decided to implement a chatbot.
>
> This chatbot will be designed to assist customers in various ways, from helping them choose the perfect product based on their hair type and concerns, to answering frequently asked questions about product usage and ingredients, and even assisting with order tracking and returns.

By posing questions and presenting scenarios, you can gain insights into user preferences, behaviors, and pain points.

Persona creation is based on hard facts and user-specific data. Using ChatGPT is not a shortcut for real research, but it can help.

Using ChatGPT for user research

It's beyond the scope of this book to outline the process of gathering information about your target audience through surveys, interviews, and market research. There is also no substitute for this. Here are some external sources to learn more about this process:

- https://www.questionpro.com/blog/research-process-steps/
- https://medium.com/@h_locke/every-ux-book-i-recommend-and-why-673d634ca530
- https://www.nngroup.com/articles/why-you-only-need-to-test-with-5-users/
- https://www.usability.gov/what-and-why/user-experience.html

However, it is feasible to use ChatGPT to help you with some ideas to carry out your research. You can do this with the following prompt:

```
Create an outline for market research, including surveys, interviews,
and additional market research methods for a luxury haircare brand:
```

The output can, at the very least, form a basis to start your research. This might look like this:

```
A. Customer Interviews:
Purpose: Gain deeper insights into the customer experience.


Sample Questions:
Describe your haircare routine.
What led you to try our brand?
Are there any challenges you have faced with our products?
B. Expert Interviews:
Purpose: Gather insights from industry experts, such as hairstylists
or dermatologists.
Sample Questions:
What trends are you currently noticing in the luxury haircare market?
How do you think our brand is perceived in the market?
IV. Additional Market Research:
A. Competitor Analysis:
Purpose: Understand the competitive landscape.
Identify key competitors in the luxury haircare market.
Analyze their product range, pricing strategy, marketing efforts, and
customer reviews. Identify their strengths and weaknesses.
B. Focus Groups:
Purpose: Gather qualitative data on potential new products or
marketing strategies.
```

```
Assemble a diverse group of individuals who are part of the target
market. Present them
with new product concepts, packaging designs, or marketing materials.
Facilitate a discussion to gather their feedback and impressions.
C. Market Trend Analysis:
Purpose: Identify current trends in the luxury haircare market.
Review industry reports and publications for data on emerging trends.
Analyze social media and influencer content to identify popular
products and ingredients.
…..
```

Once you have your details, you can also use ChatGPT to analyze the data and surface some details about the user groups:

```
Based on the following data {data}, please describe the main
demographic characteristics and preferences of our target audience.
Preferences:
Characteristics:
```

The preferences and characteristics of LuxeLocks users can then help us build up our user persona and supply steering for our chatbot persona.

Creating our user and chatbot personas

You can use the demographic characteristics created from the earlier prompt to help you create your user and chatbot personas.

Creating our user persona

Let's prompt ChatGPT to create a persona and use the results from our earlier prompt for preferences and characteristics. Replace *characteristics* and *preferences* with those values in the prompt that follows:

```
Create a user persona description for a luxury haircare brand. Users
tend to have these Characteristics: *characteristics*
Preferences: *preferences*

The persona is to be the user of our luxury haircare brand over
multiple channels. The brand sells high end haircare products which
are targeted at specific hair types:
Use the following template:
Name: {first_name}
Age: {age}
Relationship status: {relationship_status}
Occupation: {occupation}
Location: {location}
```

```
Interests: {interests}
Bio: {bio}

{first_name}'s motivations:
1. {motivation 1}
2. {motivation 2}
3. {motivation 3}
{first_name}'s goals:
1. {goal 1}
2. {goal 2}
3. {goal 3}
{first_name}'s frustrations:
1. {frustration 1}
2. {frustration 2}
3. {frustration 3}
{first_name}'s Key concern/barrier to purchase:
1. {barrier 1}
2. {barrier 2}
3. {barrier 3}
A quote from {first_name}'s: {quote}
```

We can use the resulting persona description in follow-up questions to ChatGPT.

Creating our chatbot persona

You can also use ChatGPT to create your chatbot or virtual assistant's persona. As with any prompt, you can vary the amount of information you include, but I've found that more detailed prompts lead to better results. Try using something like this:

```
Create our luxury haircare chatbot persona:
Name and Title:
What is the chatbot's name?
Does the chatbot have a title or role (e.g., Customer Care Assistant,
Haircare Expert)?
Background and Purpose:
Why was this chatbot created?
What is its primary role in interacting with customers?
```

We can now use our users' and chatbot's personas to guide ChatGPT in further tasks related to creating dialogue and interactions.

Simulating conversations

ChatGPT can be used to create sample conversations with potential users with a specific persona, which can be used to help create your conversation flows. Again, I'm stressing that this should not replace the conversational flow design process, but that it is an effective way of looking at specific use cases.

What is a sample dialogue?

Sample dialogue entails examples of specific multi-turn conversations between a user and an agent.

For those involved in the design and development of conversational experiences, these bits of dialogue are an essential tool. They allow us to explore the potential pathways a conversation can follow, and to evaluate which of these branches call for consideration and integration into the final conversational design.

```
User: What's the luggage limit for my flight to Paris?
Chatbot: May I know the airline you are flying with?
User: It's AirExcellence.
Chatbot: For AirExcellence, you're allowed one carry-on up to 10 kg
and one checked bag up to 23 kg. Carry-on size: 55 cm x 40 cm x 23 cm.
User: Thanks!
Chatbot: You're welcome! Have a great flight.
```

Let's now look at creating our own chatbot dialogue in the next section.

Creating sample dialogue

In the following prompt, we supply information for tone, language, clarification, and elements of the conversation. We want to provide these in this case to ensure we have solved the problem:

```
Generate a conversation between a user and a customer support chatbot.
The user is trying to troubleshoot a problem with their internet
connection. The chatbot should follow these content guidelines:
1. Maintain a polite and professional tone throughout the
conversation.
2. Use simple, clear language that is easy for the user to understand.
3. Ask clarifying questions to accurately diagnose the problem.
4. Provide step-by-step instructions to resolve the issue.
5. Confirm that the user's problem has been resolved before ending the
conversation.
6. If the chatbot cannot resolve the issue, it should suggest the user
to contact a human support representative.
User: "Hello, I'm having trouble with my internet connection. It's
really slow."
```

This results in a good conversational example. It's worth trying different variations so that you can build up an understanding of the sorts of ways a conversation can go.

At this stage, we're designing our flows, but we'll also see later that we can use a similar technique as part of our testing process.

Creating sample dialogue with personas

We can build on the creation of bits of sample dialogue, which have so far been centered on specific use cases, by passing more information about the user and agents' personas into the prompt:

```
We are going to create some sample dialogues between our luxury
haircare chatbot and a user:
User persona:
Name: Sophia
Age: 42
Relationship status: Single
Occupation: Corporate Wellness Coach
Location: London, UK
Interests: Pilates, Culinary Arts, Wine Tasting, Sustainable Fashion,
Volunteering
...
```

The trick here is to provide ChatGPT with details of your user and chatbot personas. Once you've provided these to ChatGPT, you can then give follow-up prompts requesting further examples of dialogue or use cases.

This technique will also work when looking at example use cases. In the next prompt, we will build on this and ask for another conversation example:

```
Using the persona descriptions of Sophia and Lila, create a
conversation between them in which Sophia is asking about the best
shampoo for her hair type.. Lila asks whether Sophie would like to do
her quiz so she can learn about her hair.
```

ChatGPT's response is impressive here, as it goes one step further and creates some quiz questions, although the best choice would be to include details of your quiz.

The technique of detailed prompting to create conversation examples can also be used in testing and further iterations, which we'll cover in the next section.

Testing and iteration in conversation design with ChatGPT

Conversation design is an iterative process. Automated agents are always changing and being improved. In this section, we'll look at testing and improving our agents with ChatGPT.

Testing with ChatGPT

In the earlier sections, we saw that ChatGPT can be a valuable tool in streamlining the design process by creating dialogue to help us craft our designs. We can carry out a similar approach when we test our agent before deployment.

We often create our agents based on which questions we think they are going to be asked. This is even more likely to be the case if we are starting a conversational AI project from scratch with little or no training data. If this is the case, it's even more important to test your agent before releasing your agent into the wild.

Even if the conversational AI project has internal testers, or if you can expose friends or family to your agent before alpha or beta releases, it's still valuable to test as thoroughly as possible before handing it over to human users. This is where ChatGPT can be utilized.

Conversation simulations for testing

This process is similar to the design process. We can create simulations where ChatGPT plays the role of a user. You can then build up sets of test user interaction scripts to follow with the chatbot without real user engagement.

We'll consider this in the context of our luxury haircare chatbot example. In the same way that ChatGPT was used to create entities and utterances, we can also use it to generate additional training data that mirrors real-world conversations.

> **Tip**
>
> If you continue with the same ChatGPT session from before, we'll already have prompted our personas. If you're starting a new session, remember to add these to your prompt to set the context.

So, we'll simply ask ChatGPT to create training data results in the form of questions that cover a range of topics related to haircare, product recommendations, sustainability, and personalized routines, reflecting Sophia's persona and interests:

```
Create 10 questions which Sophia could ask Lila
```

We can also prompt for more long-tail examples so we can test these with our agent:

```
Create 10 long tail questions which Sophia could ask Lila
```

Once we have a set of training questions, you can test our agent manually or create an automation if your chatbot provides APIs.

Creating more specific test scripts

For more complex conversational interactions, ChatGPT can also be used to create examples of the same flow so they can be used to test your agent.

Consider an example inquiry from a user interested in the status of their order. This can be used as a prompt to ChatGPT to create other examples with different conversational turns:

```
Consider the following enquiry from Sophie about her order:
####
Sophia: Hi, placed an order for the Curl Harmony Shampoo yesterday.
Can you tell me when it will be shipped?
Lila: Welcome back, Sophia! I'm glad to assist you. Your order for
the Curl Harmony Shampoo is scheduled to be shipped tomorrow. You'll
receive an email confirmation with the tracking details once it's on
its way. Is there anything else I can help you with?
Sophia:
#####
Create 3 examples of this conversation when Sophie asks follow up
questions
```

You can then run through these script examples with your chatbot to see how it performs.

Testing for failures and edge cases

Other conversational outcomes for the same use case can also be considered and accounted for with ChatGPT. For example, you can use ChatGPT to cover edge cases that need to be accounted for within the design.

Some examples could involve users straying away from the happy path, or the following example, where there are issues with the services Lila relies on to provide answers:

```
Give 1 example where Lila has technical issues so she can't access any
order details.
```

You can also prompt ChatGPT to provide its own suggestions here:

```
Provide 3 examples of what else could go wrong with the conversation
```

The results can be used to check whether you've missed anything in the design and implementation.

Iterating your chatbot with ChatGPT

As soon as you have real users testing your agent, you can use transcripts and interaction logs to see how your conversation designs can be improved by feeding examples to ChatGPT:

```
Look at the following chat transcript:
{transcript}
Provide an example of how to improve this conversation
```

This is particularly useful if you want to look at failed conversations, conversations that ended with unsatisfactory results, or conversations that ended with a dissatisfied user. It lets you look at ways you can improve your conversation flow.

Summary

In this chapter, we focused on how ChatGPT can be used in conversation design. We explored the multifaceted role of the conversation designer and looked at some of the many tasks that make up the conversation design process in more detail. We learned that some of these tasks can be tackled with the help of ChatGPT.

We delved into the practical tasks involved, such as understanding user needs, designing personas, creating conversation flows, and writing dialogue.

We also learned that it's important to understand that ChatGPT is not a replacement for good conversation design practices, but that it is an incredibly powerful tool.

At the heart of our use of ChatGPT as a conversation design assistant is the effective implementation of prompt engineering. We started to look at the principles involved with prompt engineering. In *Chapter 4*, we'll look at prompt engineering in more detail to gain a deeper understanding of this emerging field. In the next chapter, we'll look at the different ways of using ChatGPT.

Further reading

The following links are resources to help you with this chapter:

- https://chat.openai.com/
- https://www.deepset.ai/blog/the-beginners-guide-to-text-embeddings

Part 2:
Using ChatGPT, Prompt Engineering, and Exploring LangChain

This part focuses on the hands-on use of ChatGPT, the intricacies of prompt engineering, and a deep dive into LangChain. You'll explore different ways to interact with ChatGPT, whether through the web interface, APIs, or official libraries. You'll also gain insights into prompt engineering, learning how to craft effective prompts. Finally, this section introduces LangChain, guiding you through both its basic and advanced uses, including debugging techniques, memory management, and leveraging agents for enhanced functionality.

This part has the following chapters:

- *Chapter 3, ChatGPT Mastery – Unlocking Its Full Potential*
- *Chapter 4, Prompt Engineering with ChatGPT*
- *Chapter 5, Getting Started with LangChain*
- *Chapter 6, Advanced Debugging, Monitoring, and Retrieval with LangChain*

3

ChatGPT Mastery – Unlocking Its Full Potential

In this chapter, we'll delve into the technical aspects of using ChatGPT so that you can get the most out of the technology and interact with the OpenAI models effectively with several different techniques. We'll walk through the four different ways to interact with ChatGPT – that is, through the webchat interface, OpenAI Playground, directly using the **application programming interface (API)**, or by using one of the official OpenAI libraries, all of which we'll look at in more detail.

You will gain a solid understanding of the ChatGPT interface and the difference between the Free and Plus versions, as well as the power of custom instructions. Next, you'll look at how to use AI Playground. Finally, we'll cover how to interact with the API directly and look at the OpenAI Python and Node.js libraries.

By the end of this chapter, you'll have a solid understanding of the different ways to interact with ChatGPT and be able to make informed decisions on which techniques to use.

In this chapter, we're going to cover the following main topics:

- Mastering the ChatGPT interface
- Exploring OpenAI Playground
- Learning to use the ChatGPT API

Technical requirements

In this chapter, we will be using ChatGPT extensively, so you will need to sign up for a free account. If you haven't created an account, go to `https://openai.com/` and click **Get Started** at the top right of the page or go to `https://chat.openai.com`.

For the getting started examples, depending on your coding language, you will need an environment with Python installed.

Mastering the ChatGPT interface

Using the ChatGPT interface is where most users started with ChatGPT and it's the easiest way to interact with the technology. At the time of writing, you can use ChatGPT on iOS and Android and via the web.

The Free and Plus versions of ChatGPT

At the time of writing, ChatGPT comes in Free and Plus versions. The free version allows you to use ChatGPT with some restrictions, while the Plus version includes some more powerful features and access to the latest GPT-4 model.

Let's look at the differences between the two versions and their availability across devices.

- **Free version – basic ChatGPT features**:

  - **Models**: GPT-3.5

  - **Features**: No plugin support

  - **Cost**: Free

  - **Support**: No customer supports

  - **Response time**: Has lower response times compared to the Plus version

- **ChatGPT Plus – expanded features with premium access**:

  - **Models**: GPT-3.5 and GPT-4

  - **Features**: Access to beta features, including **Browsing**, **Plugins**, and **Advanced Data Analysis**

  - **Cost**: $24.00 per month

  - **Support**: Customer support

  - **Response time**: Response times are faster than the Free version, although GPT-4 currently has a cap of 50 messages every 3 hours

Should you go for the Free or Plus version?

Whether you go for the Plus version is down to your budget and whether you need the extra features. The Plus version does provide faster response times and access to the latest GPT-4 model, which makes it worthwhile. If you're looking to achieve more complex tasks, then GPT-4 gives much better output and can handle more involved tasks.

Also, because the Plus version offers beta features, you can use plugins, which dramatically extend ChatGPT's capabilities.

ChatGPT interface

In this section, we'll provide a detailed breakdown of the features you'll see when using the main ChatGPT interface while using images as references. Although largely self-explanatory, some areas call for further details. The following screenshot shows the current ChatGPT interface:

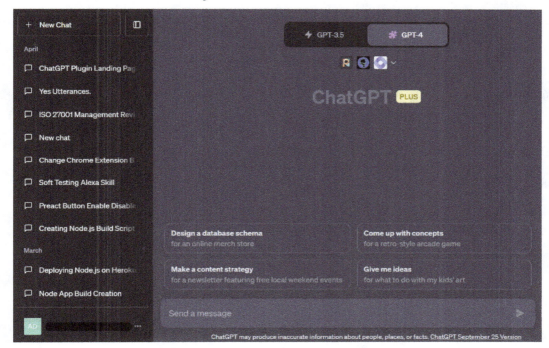

Figure 3.1 – ChatGPT interface

Once you've logged in to your OpenAI account, you can start using the ChatGPT (https://chat. openai.com/) interface. Let's become familiar with the AI tool's interface.

Sidebar

On the left part of your screen is the sidebar, which holds the following features to manage conversations and your account:

- **New Chat**: This allows you to start a new conversation. Use this button if you want to start a new dialog without the context of an earlier discussion. Remember that the ChatGPT model retains its memory of past conversations to provide contextual responses. So, if you want to change the topic of the conversation, it's wise to start a new one.

- **Chat History**: The left sidebar stores all your past dialogs, allowing you to revisit them whenever necessary. You can also turn this feature off in **Settings | Data Controls**.

- **Account Options**: Clicking on your email address or name at the bottom of the sidebar shows a popup that provides access to the following options:

  - **My Plan**
  - **Custom Instructions**
  - **Settings and Beta**

You can also upgrade if you don't have ChatGPT Plus.

Chat view

The chat screen is where you interact with ChatGPT. It's designed to mimic a messaging platform, offering a user-friendly environment where you can input your queries and receive responses from the AI:

- **Choose a model from the dropdown**: For Plus users, you can choose which GPT model you want to interact with: **GPT 3.5**, **GPT 4**, or **Plugins**.

- **Prompt examples**: There are also four buttons that you can select, along with examples of prompts to try

- **Text input area**: Here, you can type your questions and responses

- **Responses**: When ChatGPT responds to your queries, you can provide feedback by selecting the **Thumbs Up** and **Thumbs Down** buttons and copying the response to your clipboard

- **Bottom bar**: This tells you which released version of ChatGPT you are working with and includes a link to the release notes. This is useful for checking the latest features and whether they are available in your region.

Custom instructions

One issue with ChatGPT until recently is having to repeat context and instructions in prompts for every session. You had to store your prompts somewhere and remember which ones you used previously.

The **Custom instructions** feature fixes this as it allows users to personalize their interaction with the AI by providing it with specific instructions that it will remember in all future conversations.

Custom instructions are guidelines for ChatGPT to help it understand your preferences and needs without you having to include all this in your prompts. You can provide instructions covering what you would like ChatGPT to know about you so that it can provide better responses.

Let's look at a specific example for this subject matter: an instruction for a conversational AI professional.

Example of a conversation designer instruction

I am a conversation designer working on creating engaging and natural dialogs for a virtual assistant. I often need help brainstorming dialog flows, generating natural responses, and understanding best practices in conversation design.

With this instruction, ChatGPT will be aligned to assist in several ways:

- **Designing flows**: Whenever the designer asks for assistance in creating dialog flows, ChatGPT can provide structured and detailed dialog sequences that align with industry standards

- **Natural responses**: ChatGPT can help generate natural and engaging responses that can be used in the virtual assistant's script, saving time and effort for the designer and providing variation when they want to provide different variations

- **Best practices**: The designer can inquire about the best practices in conversation design, and ChatGPT will provide insights based on the latest industry standards and trends

- **Feedback and review**: The designer can ask ChatGPT to review dialogs they have created and receive constructive feedback to enhance their work

- **Resource recommendations**: ChatGPT can suggest books, courses, and other resources to help the designer further their knowledge and skills in conversation design

By setting up custom instructions, the conversation designer can have a personalized assistant that understands their professional needs and assists them efficiently in their work. It essentially becomes a collaborative tool that understands the nuances of conversation design and offers tailored assistance without having to input this into subsequent sessions.

How to set up Custom instructions

Setting up **Custom instructions** is a straightforward process and is available to both Free and ChatGPT Plus users, where they can use it on the web or via the Android and iPhone apps.

Simply click on the meatballs menu (**...**) and select the **Custom instructions** option. You will see two text boxes: one for adding details about yourself (such as age, location, hobbies, industry, and job type) and another for specifying the instructions that ChatGPT should follow (such as the tone and length of the responses).

> **Tip**
> It's important to note that custom instructions are not supported when using the API. Instead, this can be achieved by putting the custom instructions into the system prompt message.

GPTs

GPTs are custom versions of ChatGPT, built for specific tasks or purposes. You can create your own GPTs without coding by giving instructions, adding extra knowledge, and creating actions. Actions allow you to pull in information from outside ChatGPT, enabling you to call external APIs and use the returned data in your GPT's knowledge base. The Create GPT interface walks you through the process, making it easy to build your own.

Think of GPTs as specialized versions of ChatGPT, tailored for your needs. For example, you could create a GPT to help you learn the rules of a board game, teach your kids math, or design stickers. You can even share these GPTs with others in the GPT Marketplace, potentially earning money if they prove popular.

The ChatGPT community is actively building and sharing GPTs for various purposes. The "GPT Store" makes it easy to find and use these creations. GPTs represent a new way to customize ChatGPT and unlock its potential for specific applications.

At this point, you have a good understanding of how to use the ChatGPT interface. In the next section, we'll look at the second of our three methods to use ChatGPT and the other GPT models: OpenAI Playground.

Exploring OpenAI Playground

OpenAI Playground provides a free-to-use web-based sandbox environment that makes it easy for users to test and experiment with ChatGPT and the GPT series of language models. The platform also provides useful guides and examples of tasks that can be carried out with the models. Users can interact with a range of models and save their playground sessions so that they can return to them or share them with other users.

Getting started

At this stage, I'm assuming that you've created a free OpenAI account. Once logged in, you can access OpenAI Playground by clicking on the **Playground** link at the top of the page.

The following screenshot shows OpenAI Playground's landing page:

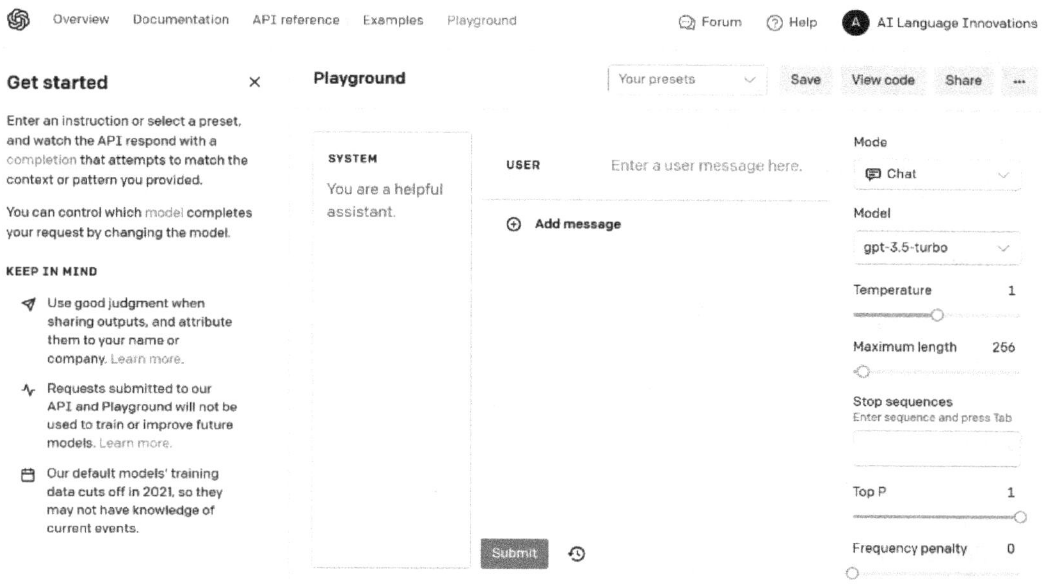

Figure 3.2 – OpenAI Playground

The main text area is where you can interact with the model. Try inputting a question using one of the prompts from earlier in this book. You can also load presets. OpenAI has dozens of pre-made prompts built in. Click on **Your Presets | Browse examples** and select one from the list. It's worth looking at some of these so that you can understand the sorts of tasks you can do and what's needed to achieve them.

UI features

OpenAI Playground is easy to grasp but let's take a quick look at the core features, all of which provide several useful features.

Saving your presets

You can save your playground's state by clicking on the **Save** button. At this point, you can toggle whether you want to make your preset accessible to anyone with the link. These presets will also be displayed in the **Presets** dropdown so that they can be used later.

Mode

The **Mode** setting allows you to decide how you want the system to respond to your prompts. It's best to leave this set to **Chat** as the other two choices are now deprecated.

AI models

You can select the model you wish to interact with by using the **Model** dropdown. The list contains all the latest GPT-3 and GPT-4 models, as well as some of the legacy versions.

Parameters

Here, you can configure the following new parameters to get different results for your prompts. I'll cover these in more detail when we cover prompt engineering:

- **Temperature**: This can be set between 0 and 1 and dictates the AI's creativity level. By default, it is set to **0.7**, which generally gives better performance for my creative tasks.

- **Maximum length**: This governs the extent of both the input prompt and the resultant output, measured in "tokens" rather than words or characters. A single token approximates four English characters.

- **Stop sequence**: This instructs the AI to cease generation at a specified point. Typically, in a conversational AI use case setup, it's useful for the model to halt after generating a one-line reply.

- **Top P**: This parameter offers another way to steer the randomness and creativity of the generated content by ranking tokens based on their relevance to the existing prompt to guide the output.

Content filter preferences

Clicking the three dots on the menu displays the option to choose the content filter preferences for the playground session. Toggling this on enables a warning if content involving sexual themes, hate speech, violence, or self-harm is found.

History

Clicking on the **History** button loads up the last 30 days of your usage. Selecting any one of your sessions allows you to view the session and gives you the option of restoring this version, which will override your current session.

Viewing code

For me, one of the key features of the playground is the ability to see the underlying code that drives the operations and outputs generated within. Click on the **View Code** option at the top of the menu and inspect the code that was executed when you were interacting with the chosen model.

Upon selecting the **Library** dropdown at the top, you can choose from the following options to see examples of how to interact with the OpenAI models:

- **API** using curl and JSON payloads
- **SDKs** using Python and Node.js

In the next section, we'll cover these approaches to using ChatGPT technology in more detail.

Pricing for API and Playground

OpenAI pricing is based on a pay-as-you-go model. Upon initial registration, users receive a premium trial with credits that are valid for the first 3 months; after this period, the credits expire.

The consumption of credits varies with each model and these costs are documented on the OpenAI website. The billing metric is based on the number of tokens utilized, with every 1K token, equating to roughly 750 words, incurring a charge.

> **Tip**
> When you're using the playground, you are still calling the same API endpoint you would use when calling the OpenAI endpoints directly. So, the playground and API calls cost the same.

How to track and control token usage

You can view your LLM usage tracking via the following dashboard:

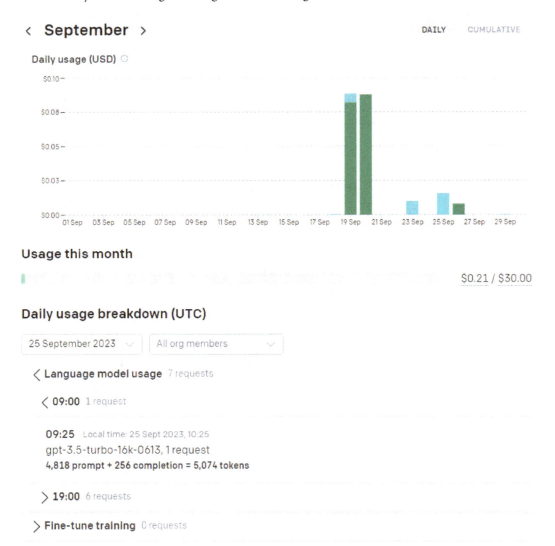

Figure 3.3 – OpenAI usage dashboard

Click on your username at the top right of the page to show the main menu and go to **Manage account | Usage.**

You can then see how many tokens you've used during the current and past billing cycles, as well as a breakdown of tokens for each request. You can also view other stats, such as how many credits you have left to spend.

For your peace of mind, it's also a good idea to set usage limits so that you can manage your spending. You can do this by accessing **Organization Billing | Usage limits**. Here, you can set hard and soft limits.

This is particularly important if you have several team members using the API. It's recommended to check the usage tracking dashboard regularly.

Learning to use the ChatGPT API

For many of the uses we covered in *Chapter 2*, it makes sense to use the ChatGPT web interface or the interactive playground.

However, OpenAI provides an extensive API that allows you to carry out several tasks with their models, including interacting with ChatGPT and their other models using code. So, if you're looking to create complex applications with ChatGPT, and I'm hoping you are, then using the API is the way to go.

In this section, we'll take a closer look at the different techniques you can use to interact with the ChatGPT API. These techniques will be used throughout the remaining chapters.

Getting started

You can easily interact with the API through standard HTTP requests from any language, via the official Python or Node.js library, or a community-maintained library. It's useful to look at examples of the code needed to interact with the API by looking at the Playground and clicking on the code option to view an example for each library or `curl` request. Alternatively, you can look at the API documentation, where you will find various examples. Authentication, model names, messages, and hyperparameters are elements you must understand when carrying out interactions with the API and libraries.

Authentication

Before interacting with the OpenAI API using any of the available means, you need to authenticate with an API key, which can be created in the console. Once logged in to the OpenAI dashboard, click on your username in the top-right corner. Select **View API Keys** from the drop-down menu. Then, click **Create new secret key** and give it a name. Make sure you write down your API key as you won't be able to see it again.

Available Models

Under **Available Models**, you can specify which model to use. In this case, it's GPT-3.5 Turbo, which is the model powering the ChatGPT interface unless you selected GPT4. At the time of writing, the following models are available for chat completion:

- `gpt-4`
- `gpt-4-0613`
- `gpt-4-32k`
- `gpt-4-32k-0613`
- `gpt-3.5-turbo`
- `gpt-3.5-turbo-0613`
- `gpt-3.5-turbo-16k`
- `gpt-3.5-turbo-16k-0613`

The Playground gives the names of the other GPT models you can use, though you can refer to the models' page for more details: `https://platform.openai.com/docs/models`.

Messages

The messages array is an array of message objects that are formed via a role and content. The role can be one of the following types:

- **System message**: System messages describe the behavior of the AI assistant for the duration of the conversation – for example, "*You are an expert Python programmer who talks like a character in the film Goodfellas.*"
- **User message**: User messages are essentially the prompts that you will be sending to the API. We'll cover examples of user messages throughout this tutorial.
- **Assistant message**: Assistant messages are previous responses in the conversation.

The first message that's sent should be a system message. Messages that follow should alternate between the user and the assistant and contain content.

> Tip
> The messages array is where you will persist past messages in a conversation for your API integration. The API will not magically remember the context of the conversation. This mechanism is used by ChatGPT under the hood.

Hyperparameters

By fine-tuning these values, you can significantly influence the model's output:

- `temperature`: This is the level of randomness of the model's output
- `max_tokens`: The maximum number of tokens to generate
- `top_p`: This controls the diversity of the generated text
- `frequency_penalty`: This affects the likelihood of tokens appearing based on their frequency in the training data
- `presence_penalty`: This affects the likelihood of tokens that indicate the "*presence*" of certain entities or concepts in the generated text

By fine-tuning these parameters, you can influence the model's output, making it more deterministic, creative, or context-aware.

Calling the API directly

The API endpoints provide other functionality, such as embeddings and fine-tuning models, along with generative tasks. However, for this book, we are only looking at the functions related to chatting with our models, which is the chat/completions endpoint.

To interact directly with the ChatGPT API, you can issue a `POST` request to `https://api.openai.com/v1/completions`.

Include an authentication header bearing the API key and any other parameters you want to send to the API:

```
curl --location 'https://api.openai.com/v1/chat/completions' \--header
'Content-Type: application/json' \--header 'Accept: application/json'
\--header 'Authorization: Bearer OPEN_AI_KEY' \
--data '{
  "model": "gpt-3.5-turbo",
  "messages": [{
    "role": "user",
    "content": "Provide 3 names for dog training chatbot"}],
  "temperature": 1,
  "top_p": 1,
  "n": 1,
  "stream": false,
  "max_tokens": 250,
  "presence_penalty": 0,
  "frequency_penalty": 0
}'
```

Responses from the API include the generated text, along with other metadata, and look like this:

```
{
  "id": "chatcmpl-84DaAYWLg69y4XUD86YbXgD4iuI5j",
  "object": "chat.completion",
  "created": 1696016214,
  "model": "gpt-3.5-turbo-0613",
  "choices": [{
    "index": 0,
    "message": {
      "role": "assistant",
      "content": "1. \"Pawsitive PupBot\"\n2. \"Canine Companion
Coach\"\n3. \"Smart Bark Assistant\""},
    "finish_reason": "stop"}],
  "usage": {
  "prompt_tokens": 17,
  "completion_tokens": 25,
  "total_tokens": 42
  }
}
```

The returned payload includes details of the chat's completion, including the response itself and usage data.

This straightforward approach provides a granular level of control over the interaction with the OpenAI models, which is ideal for understanding the API endpoint's requests and responses.

Calling the OpenAI API directly using something like `curl` is straightforward but can be annoying. It's a good idea to use an API workflow tool such as Postman to simplify calling the endpoints. There is an unofficial Postman collection that you can fork and use to simplify the process.

Now that we've covered calling the API directly, in the next section, we'll look at how we get started using the Python and Node.js libraries.

Setting up with the OpenAI Python library

Setting up your environment to use the OpenAI Python library is a straightforward process. Here are the steps to get you started:

1. First, install the library with the help of the following command:

    ```
    $ pip install openai
    ```

2. Create a Python file and add the following code to it. Import the OS and OpenAI Python packages. Then, load your key from an environment variable so that you can use it with the package:

```
import os
import openai
openai.api_key = os.getenv("OPENAI_API_KEY")
```

3. To make a call to the ChatGPT API, you need to call `openai.ChatCompletion.create()`. This is the method you must call to interact with the chat completions endpoint. The following is an example of how you would call this. To make your code more reusable, wrap it in a helper function. It will take the message you want to send and other parameters to pass on to the `create()` function and return the API's response:

```
import os
import openai
openai.api_key = os.getenv("OPENAI_API_KEY")
print(openai.api_key)

def chat_with_gpt( model, user_message,
    top_p=1,frequency_penalty=0,presence_penalty=0,
    temperature=0
):
    try:
        response = openai.ChatCompletion.create(
            model=model,
            messages=[
                {
                    "role": "system",
                    "content": "You are a helpful conversational
                        AI expert."
                },
                {"role": "user",
                    "content": user_message},
            ],
        top_p=top_p,
        frequency_penalty=frequency_penalty,
        presence_penalty=presence_penalty,
        temperature=temperature)
        return response

    except Exception as e:
        return str(e)
```

The following helper function allows you to easily customize your interactions with the ChatGPT API. You can call the function as follows and print the response:

```
response = chat_with_gpt(
    "gpt-3.5-turbo",
    "Suggest a good name for a customer support chatbot working for a
holiday company",
    top_p=0.9,
    frequency_penalty=-0.5,
    presence_penalty=0.6,
    temperature=0.5)
print(response)
```

This gives us the flexibility to send whatever hyperparameters we want and change models.

Running the program

Let's look at how to run the program:

1. **Set up your environment variable**: Make sure you have your OpenAI API key set up in your environment variables. You can do this by running the following command in your terminal (replace YOUR_API_KEY with your actual API key):

    ```
    export OPENAI_API_KEY='YOUR_API_KEY'
    ```

2. **Execute the Python script**: Save the code in a file, such as chatgpt_interaction.py, and then run the script from your terminal:

    ```
    python chatgpt_interaction.py
    ```

This setup allows you to send any message you want and adjust the parameters to fine-tune the behavior of the GPT model and compare responses between different models.

Setting up with the OpenAI Node.js library

OpenAI presents a Node.js library, authored in TypeScript, which makes it a great option for TypeScript projects as the library includes TypeScript definitions. Let's go ahead and start using the library:

1. First, install the library with the help of the following command:

    ```
    $ npm install openai
    ```

2. You should use environment variables or a secret management tool to expose your secret key. For example, you can create a `.env` file to store your environment variables (make sure you add this file to `.gitignore` to avoid committing it to your version control). You can use a package such as `dotenv` to load these environment variables in your application:

```
import * as dotenv from 'dotenv';
dotenv.config();
const mySecret = process.env['OPENAI_API_KEY']
```

Reference your OpenAI API key stored in an environment variable.

3. It also makes sense to create a helper function for greater flexibility:

```
import OpenAI from 'openai';

const openai = new OpenAI({
    apiKey: mySecret, // defaults to process.env["
        OPENAI_API_KEY"]
});

async function chatWithGPT(
    model,
    userMessage,
    topP = 1,
    frequencyPenalty = 0,
    presencePenalty = 0,
    temperature = 0
) {
    try {
        const prompt = userMessage;
        const maxTokens = 100;
        const chatResponse = \
            await openai.chat.completions.create({
            model: model,
            messages: [{
                role: "user",
                content: prompt
            }],
            temperature: temperature,
            top_p: topP,
            frequency_penalty: frequencyPenalty,
            presence_penalty: presencePenalty,
            max_tokens: maxTokens
        });
        return chatResponse;
```

```
        } catch (error) {
            return error.toString();
        }
    }
```

4. The helper function allows you to easily customize your interactions with the ChatGPT API. You can call it like so:

```
async function main() {
    try {
        const response = await chatWithGPT("gpt-3.5-turbo",
            "Hello world", 0.9, -0.5, 0.6);
        console.log("ChatGPT Response:", response.data);
    } catch (error) {
        console.error("Error:", error);
    }
}
```

The models, messages, and hyperparameters are the same ones that are used in the Python library.

Handling errors

If the library fails to establish a connection with the API or receives a non-success status code such as a 4xx or 5xx response, it will throw an exception derived from the APIError class.

In this case, it's easy to handle errors coming back from the service and you can choose to handle these as needed. In our helper function, we are looking for APIError:

```
if (error instanceof OpenAI.APIError) {
    // Do something with the APIError
    console.log(error.status); // 400
    console.log(error.name); // BadRequestError
    throw error;
} else {
    throw error;
}
```

The returned error codes are as follows:

- 400: BadRequestError

- 401: AuthenticationError

- 403: PermissionDeniedError

- 404: NotFoundError

- 422: `UnprocessableEntityError`
- 429: `RateLimitError`
- >=500: `InternalServerError`
- N/A: `APIConnectionError`

By understanding these error codes and how to handle them, you can create more robust and error-resilient applications using the OpenAI Python library.

Request and response structures

This library encompasses TypeScript definitions for all request parameters and response fields. They can be imported and used as follows:

```
const parameters: OpenAI.Chat.ChatCompletionCreateParams = {
    model: model,
    messages: [{
        role: "user",
        content: prompt
    }],
    temperature: temperature,
    top_p: topP,
    frequency_penalty: frequencyPenalty,
    presence_penalty: presencePenalty,
    max_tokens: maxTokens
};
const chatResponse: OpenAI.Chat.ChatCompletion = \
    await openai.chat.completions.create(parameters);
```

By utilizing TypeScript definitions for request parameters and response fields in the `OpenAI` library, you can ensure that your code is type-safe and aligns with the expected structures of the API. This approach simplifies the process of constructing requests and interpreting responses and you can ensure a more efficient and error-resistant development experience when interacting with the OpenAI API.

Retries and timeouts

Retries for the following errors thrown by the API are automatically retried twice by default, with a short exponential backoff:

- **Connection errors**: 408, 409, and 429
- **Internal errors**: >=500

You can use the `maxRetries` option to configure or disable this for all API calls, as follows:

```
const openai = new OpenAI({
    maxRetries: 0, // default is 2
});
```

Alternatively, you can set the following parameter on a per-call basis:

```
await openai.chat.completions.create({
    messages: [{
        role: 'user',
        content: 'How can I...?'
    }], model: 'gpt-3.5-turbo' }, {
        maxRetries: 5
});
```

Similarly, API timeouts can be configured with the `timeout` option. The default timeout is 10 minutes. You can set the default for all requests:

```
const openai = new OpenAI({
    timeout: 20 * 1000, // 20 seconds (default is 10 minutes)
});
```

You can also set the timeout on a per-request basis:

```
await openai.chat.completions.create({
    messages: [{
        role: 'user',
        content: 'How can I......?' }],
    model: 'gpt-3.5-turbo' },
    {timeout: 5 * 1000,}
);
```

It's a personal preference for what language to use. There are also several open-source libraries available.

Other ChatGPT libraries

If you want to interact with the OpenAI API using other languages, there are lots of options. The open-source community has developed libraries in various programming languages to interact with OpenAI's services. These libraries cover a range of programming languages and provide bindings and convenient methods for working with OpenAI's APIs.

If you're working with Microsoft's Azure, specific libraries are provided to interact with OpenAI services on Azure, such as the Azure OpenAI client libraries for Java and .NET, which are adaptations of OpenAI's REST APIs.

Together, these libraries form a rich ecosystem that developers can leverage to interact with OpenAI services and models, either directly or through cloud platforms such as Azure.

Summary

In this chapter, the focus has been on the various methods to interact with ChatGPT, each with its unique set of advantages, catering to different user needs and technical proficiencies. Initially, we looked at the ChatGPT interface, distinguishing between the Free and Plus versions while highlighting the latter's enhanced features, which include faster response times and access to the GPT-4 model, which is instrumental for more complex tasks. We also covered the novel feature of custom instructions, which significantly optimizes user interaction by retaining specified instructions across sessions.

Moving on to OpenAI Playground, we looked at how this sandbox is a great platform for experimentation with OpenAI models. With a user-friendly interface, it allows for hands-on exploration of models, saving and sharing sessions, and viewing the underlying code, serving as a bridge between a point-and-click and a more code-driven approach.

In the final section, we covered the different ways you can interact with the OpenAI API.

The API serves as the primary means for developers to use ChatGPT. You learned how to do this by following a detailed walkthrough of using the API directly via HTTP or by using OpenAI's Python and Node.js libraries. We explained the process of making API calls, handling errors, and configuring retries and timeouts, making it a comprehensive guide for developers.

At this point, you may be wondering when you should use the API rather than the web interface or the Playground. For casual interactions and many of the conversation design tasks we've looked at so far, it makes sense to use the hugely popular chat interface, which handles your conversation context, custom instructions, and plugin interactions well.

If you want to start looking at the other models and gain some exposure to the code behind interactions with the API, then OpenAI Playground is a good next stop.

However, if you want to include ChatGPT in your own conversational AI application, internal services, or data pipeline or you just want to have more control over your interactions with your model interactions, then the API is more appropriate. Leveraging the API through OpenAI's libraries or community-driven libraries in your team's language of choice is the way to go. Whatever your choice, at this point, you should be comfortable with all these methods.

So far, you used prompts extensively in *Chapter 2* when learning how to use ChatGPT for conversation design tasks. In the next chapter, we'll take a deeper dive into the new field of prompt engineering.

Further reading

The following links are a curated list of resources to help you with using ChatGPT:

- `https://platform.openai.com/playground`
- `https://beta.openai.com/account/usage`
- `https://platform.openai.com/docs/models/gpt-4`
- `https://platform.openai.com/docs/models/gpt-3`
- `https://community.openai.com`
- `https://platform.openai.com/docs/models/overview`
- `https://openai.com/policies/usage-policies`
- `https://github.com/openai/openai-cookbook/`
- `https://platform.openai.com/docs/api-reference/chat/create`
- `https://www.postman.com/devrel/workspace/openai/overview`

4

Prompt Engineering with ChatGPT

In this chapter, we will turn our focus to the art and science of prompt engineering for ChatGPT, a critical skill set for anyone aiming to harness the full potential of conversational AI. We'll dissect the anatomy of a prompt, exploring how its structure, tone, and complexity can influence the model's responses.

You'll learn how to craft prompts that set the right tone for professional conversations, and we'll share techniques to introduce complexity without overwhelming the model. We'll also delve into the strategic use of paragraphs and bullet points to enhance readability and comprehension.

Moreover, we'll discuss how to strike the perfect balance between providing ample context and maintaining brevity in your prompts. For those looking to adopt a persona, we'll offer insights into mimicking the expertise of a conversational AI engineer through your prompts.

By the end of this chapter, you'll be well equipped to engineer prompts that not only elicit accurate and contextually relevant responses but also elevate the capabilities of ChatGPT to new heights.

In this chapter, we'll cover the following key areas:

- Going through the concepts of prompt engineering
- Understanding the core components of a successful prompt
- Working with a prompt engineering strategy
- Knowing the prompt engineering techniques

This chapter aims to be your definitive guide to mastering prompt engineering, enabling you to interact with ChatGPT in a more effective and nuanced manner.

Technical requirements

In this chapter, we will be using ChatGPT extensively, so you will need to be signed up with a free account. If you haven't created an account, go to `https://openai.com/` and click **Get Started** at the top right of the page, or go to `https://chat.openai.com`.

To carry out the examples demonstrated in this chapter, I would advise using ChatGPT via the web or mobile app unless stated otherwise.

The last example technique requires Python 3.9 and Jupyter Notebook to be installed using the following link: `https://jupyter.org/try-jupyter/notebooks/?path=notebooks/Intro.ipynb`.

Going through the concepts of prompt engineering

With the rise of LLMs and their capabilities, prompt engineering has emerged as a new discipline, and everyone seems to be talking about it. We can leverage prompt engineering to help us tackle a broad spectrum of tasks with LLMS, from commonplace question-answering to more involved integrations with other systems. For anyone using LLMs, it's an important skill to master.

So, what actually is prompt engineering? In a nutshell, prompt engineering is creating the most optimal input to get the most out of interactions with LLMs. So, the better your prompt is structured, the better the output will be.

If you're sending vague prompts, then the reality is that you'll get vague or incorrect results back. Indeed, providing a simple online prompt will often get you some interesting results. However, they're unlikely to be the results you're hoping for, particularly if you're looking to complete more complex tasks.

Here is an example of a simple prompt. Try this out with ChatGPT:

```
Tips on conversation design
```

OK, this does return some useful information. However, we can improve on this if we want to return more targeted results:

```
"Suggest methods to enhance a conversation design to better capture
and retain user engagement, with a particular focus on utilizing
empathetic language and providing meaningful responses."
```

In this refined prompt, we provide more clarity and specificity to ensure that the AI's assistance is directed toward the area of conversation design that we're interested in.

Creating a successful prompt is not just about asking the right question but also about providing enough information to the LLM to create the right answer and ensuring everything is in the optimal format for the LLM to understand. Engineering a prompt is all about knowing which techniques to use depending on the task you're trying to achieve.

In the next section, let's start by looking at the different components of a prompt and considering other specifics, such as the best format to use.

Understanding the core components of a successful prompt

There is no catch-all for a successful prompt. Each task you want to carry out is different and so are the prompts needed to achieve them. However, there are some rules to follow to get the best results due to the way the models have been trained and the data that has been used to train them. In *Chapter 2*, we looked at ChatGPT to help with conversation design tasks. This involved providing detailed prompts, including some key components of a successful prompt. Yes, you've already done some extensive prompt engineering and used some of the following core prompt components outlined in the following list:

- **Instruction**: Specifies the action to be taken, usually initiated by entering a verb such as design or write
- **Context**: Provides additional information to guide the model, such as the environment or purpose
- **Scope**: Narrows down the focus of the task, specifying the types of inquiries or topics to be covered
- **Role**: Defines the capacity in which the AI should operate, such as an expert or a customer service representative
- **Audience**: Indicates the knowledge level and interests of the intended recipients of the output
- **Input data**: Specifies the data to be used in the task, often separated by delimiters
- **Output data**: Details the expected format and structure of the output, guiding the AI's response

The following figure illustrates the core components that contribute to crafting a successful prompt:

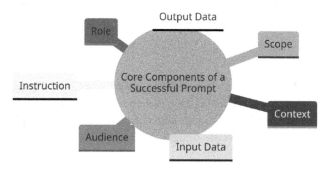

Figure 4.1 – The core components of a successful prompt

Let's look at each of the key components for a successful prompt in more detail.

Instruction

Give explicit instructions for your prompt. It's good practice to start a prompt with a verb, such as design, draft, create, generate, write, or produce. This way, ChatGPT should understand exactly what you want it to respond with and leave no room for misinterpretation. Also, tell ChatGPT exactly what you're expecting to see in a response, such as a conversation flow, website copy, blog post, chat transcript, or description.

Let's look at an example prompt to ask ChatGPT to give us a chatbot conversation example:

```
"Design a conversation flow for a customer service chatbot that
handles product returns and exchanges, focusing on a friendly and
efficient user experience"
```

In this prompt, we give a precise instruction: design a conversation flow. So, the clearer and more precise the instructions are, the less you leave ChatGPT to guess what you are trying to accomplish.

Context

To improve on a prompt, we can supply more information beyond the first instruction. Let's continue the example by providing some context to guide the model in understanding the need and purpose of our conversation design request:

```
"Design a conversation flow for a customer service chatbot"
```

We've added for a customer service chatbot to our prompt. This should help ChatGPT tailor the dialogue and actions of the chatbot. So, with extra context, we can guide the model to understand the environment where the conversation flow will be used.

Scope

Like any project brief or statement of work, it makes sense to provide ChatGPT with more detail about the scope of the task. In our case, our conversation design is for a customer service chatbot, which could be any type of request, from FAQs to complaints or returns. So, let's supply a scope so we can zoom in on the types of inquiries that we are designing for:

```
"Design a conversation flow for a customer service chatbot that
handles product returns and exchanges"
```

Hopefully, by providing ChatGPT with `handles product returns and exchanges`, we can ensure that the model focuses on generating conversation flows related to product returns and exchanges, avoiding off-topic diversions.

Role

One of the most important components of a successful prompt is the role component. Providing a role for ChatGPT to act in specifying the capacity in which the AI should work while generating its response. For example, you might ask it to act as an expert or with a specific point of view or capability:

```
Act as a first line customer service representative and provide
assistance on how to troubleshoot technical problems with the SAAS.
Please ensure a professional tone and intermediate level of complexity
in your response.
```

Including this role as part of your prompt will steer the language and behavior of a ChatGPT-powered customer support chatbot. When providing a role, it's important to be specific, clear, and relevant. Remember that we looked at providing more complex role details in *Chapter 2*, under section *Creating our user and chatbot personas*.

Depending on your task, the absence of a defined role in your prompt can result in a generic or vague response that lacks the desired focus or professionalism.

Audience

Specifying audience details in your prompt significantly tailors the output. For instance, consider the following prompt:

```
"Explain the principles of machine learning to a technical audience."
```

This is likely to result in an in-depth explanation employing technical terminology and discussing algorithms, data training, validation, and the underlying mathematical principles.

If you wanted to create content for a more general audience, you could try something like the following:

```
Explain the principles of machine learning to a general audience.
```

The expected output would be a simplified explanation using layman's terms and avoiding technical jargon.

By incorporating audience specifics in your prompt, you can enable ChatGPT to align its response with the knowledge level and interests of the intended recipients.

Input data

The input data is often the focus of your prompt. This can be text or data in other formats.

Put instructions at the beginning of the prompt. If you are including specific data for ChatGPT to use as part of your prompt, it's sound practice to use some sort of delimiter. I find that using ### or " " " to separate the instruction and input data works well:

```
Summarize the conversation delimited with triple hashes

Conversation: ###{conversation}###
```

Here is an example of this prompt providing our input conversation data as plain text:

```
Summarize the conversation below as a bullet point list of the most
important points. The conversation is delimited with triple hashes

Conversation: ###
User: Hi, I'd like to return a product I purchased last week.
Chatbot: Of course, I can aid with that. May I have your order number,
please?
User: Sure, it's 12345.
Chatbot: Thank you. I found your order. May I ask the reason for the
return?
User: The product has a defect and doesn't work properly.
Chatbot: I apologize for the inconvenience. I can process the return
for you. Would you like a refund or an exchange?
User: I'd prefer an exchange.
Chatbot: Great! I've started the exchange process. You'll receive an
email with further instructions shortly.
User: Thank you!
###
```

ChatGPT is adept at analyzing various input types, such as **JavaScript Object Notation** (**JSON**), **Hypertext markup language** (**HTML**), **Extensible Markup Language** (**XML**), images, markup, and plain text, among others. For instance, if your conversation data was stored as JSON, you could simply pass this into your prompt:

```
Summarize the conversation below as a bullet point list of the most
important points. The conversation is delimited with triple hashes

Conversation: ###
{
"conversation": [
  {
    "speaker": "User",
```

```
      "message": "Hi, I'd like to return a product I purchased last
week."
   }, {
      "speaker": "Chatbot",
      "message": "Of course, I can assist with that. May I have your
order number, please?"
   }, {
      "speaker": "User",
      "message": "Sure, it's 12345."
   }]
}
###
```

Ensuring consistency in the format and structure of input data is crucial for achieving reliable responses.

Once you've provided the correct input, it's important to tell ChatGPT what it's supposed to provide as output. Let's look at this next.

Output data

Specifying the expected output data and level of detail guides the AI toward generating responses that align with your requirements. We can supply detailed instructions on what we want as our output, whether it's text, formatted data, or code.

To get the best results, it's prudent to explicitly define the structure and format of the desired output. This could entail specifying the use of arrays, objects, or other data structures to organize the information. It could also involve defining the naming conventions for properties or keys, ensuring a standardized and predictable output, which in turn simplifies data extraction and further processing if you plan to use your responses in software or a pipeline. You will have applied this practice extensively in many of the examples in *Chapter 2* when we asked ChatGPT to return JSON data.

Simple text

Let's look at some more examples. In this simple example, we are asking for text output in a bullet point list:

```
Summarize the conversation delimited with ### as a bullet point
list of the most important points which should be no more than 150
characters each.

Conversation: ###{conversation}###
```

Output specific format or code

Another trick is to specify a numbered table, for example, by using the following:

```
Please can you create a numbered table of the components of a
successful chatbot for a technical audience.
```

You can then refer to the numbered items with follow-up questions:

```
can you provide more detail on number 3
```

We can also ask ChatGPT to return a more complex specific format such as tabular data in HTML:

```
Create a table summarizing the chatbot's performance across four
metrics (User Engagement, Response Accuracy, Processing Time, and User
Satisfaction) over a span of three months. Output the table as HTML.
The metrics are as follows:
- User Engagement: Jan: 1200 interactions, Feb: 1300 interactions,
Mar: 1400 interactions
- Response Accuracy: Jan: 85%, Feb: 87%, Mar: 90%
- Processing Time: Jan: 1.2 seconds, Feb: 1.1 seconds, Mar: 1.0 second
- User Satisfaction: Jan: 80%, Feb: 82%, Mar: 85%
Include a column for the average across the three months for each
metric.
```

You can also go one step further and ask for a specific format and guidance on what would be the best output based on our chosen format. In the next example, we will ask ChatGPT to provide Vega-Lite (high-level grammar for visual analytics):

```
The following is data delimited in ### summarizing a chatbot's
performance across four metrics (User Engagement, Response Accuracy,
Processing Time, and User Satisfaction) over a span of three months.
Please can you create this for vega lite representation for this data?
Include a column for the average across the three months for each
metric.
###
- User Engagement: Jan: 1200 interactions, Feb: 1300 interactions,
Mar: 1400 interactions
- Response Accuracy: Jan: 85%, Feb: 87%, Mar: 90%
- Processing Time: Jan: 1.2 seconds, Feb: 1.1 seconds, Mar: 1.0 second
- User Satisfaction: Jan: 80%, Feb: 82%, Mar: 85%
###
```

Another good example of specifying output is instructing ChatGPT to look at unstructured data and transform it into an output that we can use in further pipelines or software applications. In the following example, we will ask for JSON and specify the properties we want to be created, as well as add another property to populate with sentiment:

```
Please look at these live chat transcripts and create JSON for this
conversation as a list of message objects, add one extra property
```

```
which will include sentiment for each message:
[2023-10-09 08:00:00] User: What's the weather like today?
[2023-10-09 08:00:05] Virtual Assistant: Today's forecast is cloudy
with a high of 68°F and a low of 52°F. There's a 60% chance of rain in
the afternoon.
[2023-10-09 08:01:00] User: Do you have any suggestions for indoor
activities?
[2023-10-09 08:01:05] Virtual Assistant: Certainly! You might consider
reading a good book or visiting a local museum. Both activities are
excellent ways to spend a rainy day indoors.
```

Output of LLMs can be inconsistent and results such as lists can be brittle, so it's often advisable to return JSON if you want to use the output in software or pipelines.

As you have seen, outputs can be practically anything and it makes sense to take care of specifying exactly what we would like ChatGPT to deliver.

Conclusion

A prompt doesn't require all these elements, and the format varies based on the specific task you want to achieve. It's also worth noting that the same prompt components will work with any LLM.

While these components are not exhaustive, they should allow you to get consistently better results.

So, now that you know what the key parts of a prompt are, let's look at prompt engineering in more detail by considering your prompting strategy.

Working with a prompt engineering strategy

Prompt engineering is an iterative process. It's unlikely that you'll be successful with your task with your first, second, or third prompt. This is why I tend to avoid guides and lists on the internet offering the 100 best prompts. There are no hard and fast rules in prompt engineering, but there are some strategies to follow alongside using some or all of our effective prompt components. Let's take a closer look at some of these strategies in more detail.

Define clear goals

Before you even start using prompts, it's best to consider the task you're trying to achieve. A clear design and goal will mean fewer iterations overall. So, consider your task and see whether you can explain it clearly to yourself or describe it in a few sentences, including some appropriate keywords, to ensure that you've thought it through and have enough detail to provide to ChatGPT. LLMs tend to go off on tangents if you don't provide enough detail in your prompt, as there will be too many possibilities for them to go the wrong way.

Employ iterative prompt development

Prompt development is an iterative process. It's unlikely that the first prompt you use to explain your task to an LLM will be the prompt you use in production. What's key is that you start simple and then, with every output, iterate on the prompt depending on where the result is lacking or needs refinement. In the next example, we will illustrate iterating on the prompts for a specific task.

Iterative development example

In this example, you are tasked with designing a conversational flow for a new chatbot that will be deployed on a university's website to answer queries related to admissions, programs, and campus facilities. The conversational flow should be intuitive, informative, and engaging to assist prospective students in finding the information they need.

You have some data about the university's admissions, which can be supplied as part of your prompt:

```
Data:
Admissions:
Application Period: 6 September to 15 January
Requirements:
A-Level qualifications or equivalent (specific grade requirements may
vary by program)
Personal Statement (4000 characters max)
Academic References (1 required, 2 recommended)
Transcripts from previous institutions
```

Our first try is fairly simple:

```
Create a chatbot conversation flow for university inquiries
Data: {data}
```

This is pretty vague and lacks context and guidance, leaving ChatGPT without the necessary information to generate a meaningful conversational flow. The output from ChatGPT is pretty impressive and includes an example conversation, but it's fairly basic.

Let's try and improve our prompt with more detail:

```
You are tasked with designing a conversational flow for a chatbot
deployed on a university's website. The chatbot should be able to
answer queries regarding admissions, programs, and campus facilities
based on the provided data. Draft a conversational script for the
following user inquiry: "Tell me about the admission process and
deadlines."
Data: {data}
```

This is a better prompt; we have provided more scope and clearer instructions, as well as a scenario, and requested a script for a specific user inquiry. However, it doesn't provide any steering on the chatbot's persona or emphasis on the chatbot's tone.

```
Envision yourself as a chatbot developer crafting a conversational
flow for a university chatbot called Unipal. Utilizing the given
data, ensure the chatbot provides informative, clear, and engaging
responses regarding admissions. Here's a scenario: A prospective
student inquires, "I am interested in the Engineering program. Can
you guide me through the admission requirements and process?" Create
a conversational script that demonstrates how the chatbot would guide
the user through this inquiry, providing all necessary details in a
friendly and supportive manner.
Data: {data}
```

This prompt creates a persona for ChatGPT, places it in a realistic scenario, and provides a specific user inquiry to respond to. It emphasizes the importance of a friendly, supportive, and engaging tone, and it gives ChatGPT the context and data needed to craft a meaningful, informative response to fulfill our conversation flow task.

As you can see from this example, prompt engineering is a process of iteration. We started simply and, through a series of prompt improvements, built up the detail of our prompt based on each prompt result.

Start simple

It's also useful to keep in mind that, depending on the task you want to achieve, it can be the case that using several simpler prompts will result in better outcomes than one long complex prompt. If you are using ChatGPT with the web or mobile chat interface, you can start with a simple explicit prompt and build on this with further follow-up prompts. Remember that the ChatGPT application manages the context and conversation history behind the scenes, which you will have to implement yourself if you are using the API.

> **Tip**
>
> If you are trying to achieve too much in a prompt, LLMs can become confused and may not provide the best results. If you are trying to do more than four or five things at a time, it could be time to split these down into a series of follow-up prompts.

If you are trying to achieve a very complex task, it can be easier to not try and create the perfect prompt but to look at achieving your task in multiple prompts. Start with a clear and concise prompt and then follow up with further questions or clarifications.

Let's consider a more complex task: designing a multi-functional conversational AI system for a healthcare provider that can handle appointment scheduling, prescribe refills, and provide general information about medical conditions. Instead of asking ChatGPT to provide all the information in one go, you could start with a clear and concise prompt to lay down the foundations of the system:

```
Design a conversational interface for a healthcare provider that can
greet users and ask for their primary concern.
```

Once you have a basic structure, you can delve deeper into each functionality with follow-up questions, such as asking about further details for the appointment scheduling:

```
Elaborate on how the conversational AI should handle appointment
scheduling including checking doctor availability and confirming
appointment details with the user.
```

What you are doing here is starting simple, so ChatGPT understands the overall task, and then breaking down the complex task into smaller and more manageable subtasks. This approach works well for larger, more complex tasks such as coding and text use cases.

Use follow-up prompts to test against multiple examples

Sometimes you will have multiple examples of data to evaluate your prompt against. It makes sense to test the prompts across a range of inputs so you can see whether your prompt is performing correctly across several examples. This is particularly important if you are using ChatGPT to process large amounts of data. You can use ChatGPT to help with this. Here is an example. Let's say that you have an LLM automation that creates hotel descriptions based on a set of hotel data. This prompt could look something like this:

```
Act as a travel agent and provide a comprehensive hotel description of
200 words maximum for the following hotel data: {
  "hotel_name": "Mountain Lodge",
  "location": "Swiss Alps",
  "rating": 4.5,
  "no_of_rooms": 80,
  "facilities": [
    "Ski-in/Ski-out",
    "Fireplace",
    "Restaurant"
    ],
  "price_range": "£200-600",
  "description": "A lodge nestled among the snow-capped Swiss Alps."
  }
Output the description as a description property on the same JSON
```

It may be that you're not 100% sure that the facilities data is going to be available for every hotel, and you don't want any made-up facilities details or mention of no facilities, so you can check that the description does not mention facilities in another prompt:

```
Look at the following hotel_description and return are_facilities_
mentioned = true if facilities are mentioned or are_facilities_
mentioned = false if not.
You don't need to explain your reasoning
hotel description: "situated in the..."
```

You can see the value in testing your prompts across a range of inputs and sanitizing the output from ChatGPT with follow-up prompts here.

Use temperature when you need to

We covered how to change the temperature in *Chapter 3*. Temperature allows you to change the randomness of the model's responses. The higher the temperature, the more varied and random the response will be. Generally, when building applications where you want a predictable response, it's a good practice to use a temperature of zero. So, if you are trying to build a system that is reliable and predictable, I would recommend using zero. For applications where you would want more creativity, it's a good idea to increase the temperature.

Handling memory limitations in ChatGPT

Memory limitations in ChatGPT, or any LLM, can be a bottleneck when you're trying to maintain a long or complex conversation or understand a broader context. All the GPT models have a token limit, which includes both the input and output tokens. Once the token limit is reached, you either have to truncate, omit, or otherwise manage the conversation history to make room for new input and output.

Strategies for managing memory issues

In addressing the limitations of memory in conversational AI, several strategies can be employed to optimize the interaction flow. Here are some effective techniques:

- **Selective truncation**: Keep only essential parts of the conversation. For example, remove greetings or less relevant exchanges to save space in follow-up messages.

- **Summarization**: Periodically, you can summarize what has been talked about so far and use this in further prompts.

- **Pagination**: Break down the conversation or contextual content into pages or sections and deal with each of them independently. You can use a prompts splitter such as https://www.chatsplitter.com/ to divide longer input documents into smaller, manageable segments. This is useful if you're dealing with larger documents. This tool allows you to easily upload multiple files and have them automatically split into multiple chunks, which you can load into ChatGPT.

- **Prompt minimization**: There is some scope to minimize prompt length, but as you've learned to create optimum prompts throughout this chapter, it doesn't make too much sense to limit them. The results are probably negligible. That being said, depending on your use case, it may be worth removing data that you've added as input in your prompts.

The reset mechanism

When you've reached the memory limit, it's important to reset the model. Failing to do so can truncate inputs or outputs unpredictably, causing loss of context or incorrect answers. Resetting should be a controlled operation that keeps the essence of the conversation while discarding expendable parts. The most effective way to do this is to summarize the conversation and use it in your next prompt.

Be cautious about replicating incorrect turns from earlier iterations. If a response is off track, make sure you don't include that part when you reset the model.

The following are some different ways to reset a conversation:

- **Hard reset**: Completely clear the conversation and begin with a new initializing context
- **Soft reset**: Retain key points or summaries of the prior conversation as initializing context
- **Context caching**: Store essential information in an external database or context manager and reintroduce it as needed

Use the GPT-4 model, which has larger memory

This is an easy fix and makes significant inroads into alleviating the memory problem. The latest GPT-4 models have a much larger memory. The token limit for GPT-35-Turbo is 4,096 tokens, while the token limits for GPT-4 and GPT-4-32k are 8,192 and 32,768 respectively, so it's often wise to use these. Just keep API costs in mind.

Conclusion

As you've learned, it's important to have a solid prompt engineering strategy so you can overcome the challenges of prompt engineering. Follow an iterative strategy and ensure you provide specific, clear prompts with plenty of background information and context, and avoid ambiguity. In the next section, we'll look at some specific prompt engineering techniques.

Knowing the prompt engineering techniques

Let's use what we've learned with some ChatGPT prompting techniques and use cases that are specific to conversational AI.

In this section, we will delve into advanced prompt engineering techniques, aiming to leverage the full potential of ChatGPT for specialized use cases within conversational AI. We'll explore methods such as few-shot learning for targeted responses, summarization for data interpretation, and the iterative refinement of prompts for a travel chatbot. Each technique is designed to refine the AI's understanding and output, ensuring that the resulting interactions are as effective and relevant as possible to the task at hand. Whether it's shaping a customer support chatbot or integrating LLMs for data summarization, this section should help you learn to use the most useful prompt techniques.

Few-shot learning for a customer support chatbot

A common way to prompt ChatGPT with context and give tighter boundaries on output is to use a few-shot learning approach. In few-shot learning, a set of training examples is provided as part of your prompt. The benefit of this is that you are providing extra context, and each response gives ChatGPT examples of the output that you are looking for, so you can have a fair degree of confidence about the responses.

To look at this in action, let's consider asking ChatGPT to take on the role of a customer support chatbot for a smartphone company.

A single-turn prompt could be something like the following:

```
Act as a dog training chatbot designed to answer questions about dog
training.
- Only answer questions related to dog training
- If you're unsure of an answer, you can say "I don't know" or "I'm
not sure" and recommend  a source on the internet
```

The output from ChatGPT works well but is a little too detailed for our use case. You could give more output guidelines, or, with a few-shot learning approach, you can guide ChatGPT to answer in a more structured and consistent manner:

```
Act as a dog training chatbot called Fido designed to answer questions
about dog training.
- Only answer questions related to dog training
- If you're unsure of an answer, you can say "I don't know" or "I'm
not sure" and recommend  a source on the internet
Example conversations:
Example 1:
User: How do I stop my dog from barking at strangers?
Assistant: To curb your dog's barking at strangers, work on
socialization and desensitization techniques. Reward calm behavior and
consider using a 'quiet' command.
Example 2:
User: What's the best way to potty train my puppy?
Assistant: Consistency is key for potty training. Take your puppy
outside frequently, especially after meals or playtime, and reward
them for doing their business outside.
Example 3:
```

```
User: My dog pulls on the leash. What should I do?
Assistant: To correct leash-pulling, use a short leash and stop
walking when your dog pulls. Only resume walking when the leash is
slack. This teaches them that pulling gets them nowhere.
```

For the same example, if you were interacting with the ChatGPT completions API, your examples would be part of the array of messages you send to the API:

```
{"role": "system", "content": "You are a dog training chatbot called
Fido designed to answer questions about dog training.  Only answer
questions related to dog training - If you're unsure of an answer, you
can say "I don't know" or "I'm not sure"},
{"role": "user", "content": " How do I stop my dog from barking at
strangers "},
{"role": "assistant", "content": " To curb your dog's barking at
strangers, work on socialization and desensitization techniques.
Reward calm behavior and consider using a 'quiet' command."},
{"role": "user", "content": " What's the best way to potty train my
puppy?"},
{"role": "assistant", "content": " Consistency is key for potty
training. Take your puppy outside frequently, especially after meals
or playtime, and reward them for doing their business outside "}
```

By providing examples to ChatGPT, you're priming the model with domain-specific knowledge, as well as setting up the interactive role of the task, leading response format, and task understanding. You are hopefully also helping to maintain uniform responses.

For this use case, few-shot learning enables ChatGPT to act as a specialized assistant, transforming it from a general-purpose chatbot to one capable of supplying domain-specific guidance, for dog training in this case. In the next section, let's look at using ChatGPT to carry out another common use case with some added complexity.

Prompting to summarize data for a conversational agent

One notable use of LLMs is leveraging them for summarization tasks. Our prompt summarization example differs from a standard text summarization task. Instead, we'll be creating a prompt to answer questions about data in JSON format.

This technique is valuable if you are aiming for your chatbot or automated assistant to address inquiries about specific data for the inquirer.

For this example, suppose you have a customer support chatbot for an online travel agency. Your chatbot is handling an inquiry from a customer about their flight luggage details. The chatbot has a direct integration with the agency's back-office web service. It also has the current user's details, so it's straightforward to look up the customer's booking data. The complexity comes with interpreting and representing the data for the user to answer their question.

For our example, the flight data that is returned by the system looks like the following:

```
"flights": [{
  "status": "CONFIRMED",
  "supplierName": "Easyjet Flight",
  "segments": [{
    "direction": "OUTBOUND",
    "legs": [{
    "flightNumber": "EZY8703",
    "airline": {
      "bagsConfiguration": {
      "cabinBagWeight": "NO_LIMIT",
      "holdWeight": "20kg",
      "baggageIncluded": false,           "cabinBagDimensions": "36x45x20
cm",                                      "cabinBagAllowance": 1}
    },"carrier": {"carrierCode": "EZY","name":"easyJet"}
  }
]},{
  "legs": [{
    "airline": {
    "bagsConfiguration": {
      "holdWeight": "20kg",
      "cabinBagDimensions": "36x45x20cm",
      "cabinBagAllowance": 1,
      lugageIncluded": false,
      "cabinBagWeight": "NO_LIMIT"
      }
},"carrier": {"name": "easyJet","carrierCode": "EZY"},
  "flightNumber": "EZY8706"
    }
  ],
  direction": "INBOUND"}
  ]
}]
```

In an intent-based conversational AI system, we would construct a response to the user based on the data by parsing this data and constructing it manually or with a templating system. This can become brittle and hard to maintain. Instead, we can use ChatGPT to handle answering the question with a summary of the flight's data.

Let's look at a prompt to achieve this:

```
You are a customer support chatbot for an online travel agency. look
at the following flights information represented in JSON format
delimited with ###, use this to create a summary of inbound and
outbound flight baggage details
Summarize outbound then inbound and include all flights even if they
have no baggage
If a flight has no baggage state "My records show no confirmed hold
luggage for this part of your trip.
Include the following statement as the last part of the summary:
"Hold luggage is not per person, but for the entire booking."
flights: ###{flights}###
```

After receiving a coherent output from ChatGPT, we can iteratively refine our prompt for our specific use case. For example, we can do this by altering the output to generate a summary tailored for an IVR system:

```
create a summary of inbound and outbound flight baggage details for an
IVR system
```

As you can see, the technique of using ChatGPT to summarize works well when the input is structured data as well as unstructured text. This is also a great technique to use if you are working with an intent-based system, but you want to introduce LLM technology carefully and only at an intent level or specific to some intents.

ChatGPT can still do a lot more. Let's consider building on our example to create a complete chatbot in the example in the next section.

Prompting to create your own chatbot powered by ChatGPT

We can build on the previous example of our travel assistant agent by using ChatGPT to create our own working travel assistant chatbot.

This code example demonstrates a simple interactive chatbot created using Python, ChatGPT, and IPython widgets for the display elements within a Jupyter Notebook environment. The chatbot has some holiday information in JSON format, which is included in the context of your prompt.

Importing the necessary libraries

First, we need to import all the necessary libraries:

```
import os
import json
import openai
import ipywidgets as widgets
```

```
from IPython.display import display, clear_output, Markdown

openai.api_key = os.getenv("OPENAI_API_KEY")
```

Essential libraries and modules are imported to support the chatbot's functionality. For instance, `openai` is used to interact with the GPT-3.5-Turbo model, and `ipywidgets` and `IPython.display` are used to create an interactive UI within Jupyter Notebook. We are also passing our environment variable populated with our OpenAI key.

Loading booking data for the chatbot

Next, we need to load the chatbot booking data from the JSON file:

```
with open('booking.json', 'r') as f:
    booking = json.load(f)
```

The code reads a `booking.json` file to get booking details for an imaginary customer so your chatbot will have some information when you ask it about your holiday. This simulates an integration with a back-office system.

Defining the helper function for calling the completions endpoint

The following code outlines the helper function for calling the completions endpoint with our messages and other parameters:

```
def chat_with_gpt(messages, model="gpt-3.5-turbo",
    top_p=1, frequency_penalty=0,
    presence_penalty=0,temperature=0
):
    try:
        response = openai.ChatCompletion.create(
            model=model,
            messages=messages,
            top_p=top_p,
            frequency_penalty=frequency_penalty,
            presence_penalty=presence_penalty,
            temperature=temperature
        )
        return response.choices[0].message["content"]
    except Exception as e:
        return str(e)
```

This function sends the conversation history (`messages`) to the GPT-3.5-Turbo model and returns the model's response.

Creating UI elements

The following elements are simple UI components to make up our chat interface:

```
inp = widgets.Text(value="Hi",
    placeholder='Enter text here…')
chat_button = widgets.Button(description="Chat!")
output = widgets.Output()
panels = []
```

UI components are created using `ipywidgets`. These include a text input field (`inp`), a button (`chat_button`), and an output area (`output`) to display the conversation, as well as the panels list, which is initialized to hold the conversation panels that will be displayed.

Initializing the context

Next, we will create the context for our conversation with our system prompt and booking information:

```
system_content = f"You are Shelley an automated travel assistant to
answer questions about a customers holiday, \
start by greeting them by their first name and asking them how you can
help with their holiday \
mention where they are going \
This customers holiday details are here: \
{booking}"
context = [{'role':'system', 'content':system_content}]
```

The context array is initialized with a system message instructing the assistant with the prompt details and customer data input. This is important, as it sets the scenario for the conversation.

Defining the collect_messages function and UI elements

We will also create a function to format the messages and create our UI elements:

```
def collect_messages(change):
    # ...
    context.append({'role':'user', 'content': f"{prompt}"})
    # ...
    context.append({'role':'assistant', 'content': f"{response}"})
    # ...
```

This function is triggered when the user clicks the **Chat!** button.

It reads the user's input, clears the input field, updates the context array with the user's message, and retrieves the assistant's response by calling `chat_with_gpt`.

The context array is updated again with the assistant's response, thus forming the conversation history.

User and assistant messages are formatted and added to the panels list, which holds the conversation.

The `collect_messages` function is also bound to the click event for `chat_button`:

```
chat_button.on_click(collect_messages)
dashboard = widgets.VBox([inp, chat_button, output])
display(dashboard)
```

A vertical box (`VBox`) widget is created to hold the text input, button, and output area, forming the chatbot dashboard.

The `display` function is called to render the dashboard in Jupyter Notebook.

The chatbot context

The context array plays a crucial role in managing the conversation history and ensuring coherent and contextually relevant responses from ChatGPT. It gets updated with each exchange between the user and the assistant, providing the necessary context for generating meaningful responses based on the ongoing conversation and the initial system instruction. In essence, this is what is occurring behind the scenes when you interact with the ChatGPT UI.

Once you've run the code, you should be able to see a simple chatbot interface:

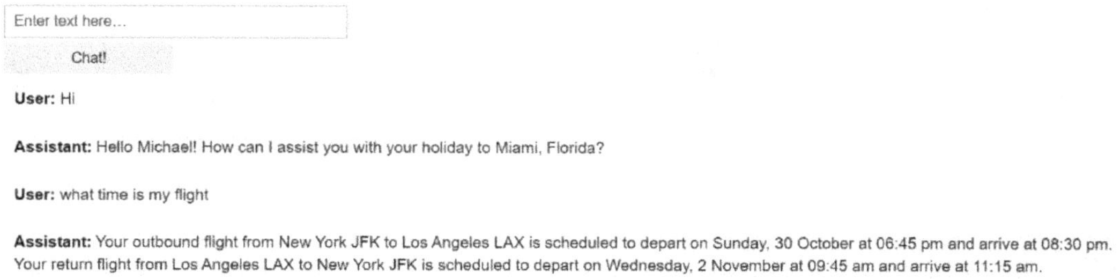

Enter text here...

Chat!

User: Hi

Assistant: Hello Michael! How can I assist you with your holiday to Miami, Florida?

User: what time is my flight

Assistant: Your outbound flight from New York JFK to Los Angeles LAX is scheduled to depart on Sunday, 30 October at 06:45 pm and arrive at 08:30 pm. Your return flight from Los Angeles LAX to New York JFK is scheduled to depart on Wednesday, 2 November at 09:45 am and arrive at 11:15 am.

Figure 4.2 – Chatbot interface

You can now start the conversation by saying hello. You should be greeted by ChatGPT, and you will be able to ask follow-up questions about your holiday.

Try changing the information in the JSON and asking other questions on your own.

You can also try changing the prompt and data input and creating a different type of chatbot and use case.

Summary

In this chapter, we looked at the emerging field of prompt engineering. You gained an understanding of the core components of a successful prompt. We introduced the importance of clearly defining roles, audience, input, and output data in prompts to tailor the model's responses to specific tasks or user groups.

With a heavy emphasis on the iterative nature of prompt engineering, we saw that it's important to have a solid prompt engineering strategy. We hopefully covered some solid details on what to look for or consider during prompt engineering.

We also looked at some prompt engineering techniques and how to use these in real-world examples specifically tailored to the world of conversational AI.

Through iterative development and testing across multiple data examples, you were encouraged to refine prompts to achieve better, more consistent outcomes from ChatGPT, so you could enhance the model's utility in various applications.

So, you should now be comfortable with moving on to more complex ChatGPT use cases. In our next chapter, we are going to delve into LangChain, an open-source framework that helps with the integration of ChatGPT with other external components, enabling the creation of more advanced applications.

Further reading

The following links are resources to help in this chapter:

- `https://www.chatsplitter.com/`
- `https://jupyter.org/`

5

Getting Started with LangChain

In this chapter, we're going to look at LangChain and how this framework enables users to build complex LLM applications. You'll gain an understanding of LangChain and the problems it fixes, as well as an introduction to the core components of LangChain, LangChain Expression Language, and the different types of chains that you can create.

At its core, LangChain is an open-source development framework designed for building LLM applications. By the end of this chapter, you'll be well-equipped to engineer LangChain applications and use the more advanced functions and capabilities covered in the next chapter.

In this chapter, we'll cover the following key areas:

- Introduction to LangChain
- Core components of LangChain
- Understanding LangChain Expression Language
- Creating different LangChain chains

This chapter aims to be an introduction to the core concepts to give you a solid base for mastering LangChain, enabling you to move on to more complex LangChain applications.

Technical requirements

In this chapter, we will be using ChatGPT extensively, so you will need to be signed up with a free account. If you haven't created an account, go to `https://openai.com/` and click **Get Started** at the top right of the page, or go to `https://chat.openai.com`.

To carry out the examples, I would advise using ChatGPT via the web or mobile app unless stated otherwise.

The last example technique requires Python 3.9 and Jupyter Notebook to be installed (`https://jupyter.org/try-jupyter/lab/index.html`).

Introduction to LangChain

As we progress further into the realm of LLM application development, you'll start to notice that these sorts of applications quickly become complex and necessitate increasingly complex prompts. A prompt can include basic system instruction, user input, subject metadata, personalization for the user, previous interaction data, and other information from other systems. So, a typical application might necessitate multiple promptings of an LLM along with parsing its output, thus calling for a substantial amount of auxiliary code. This is where LangChain, created by Harrison Chase, comes into play to add structure to your LLM applications and ease the development journey.

LangChain emerged from the necessity to fabricate intricate LLM applications while employing common abstractions in their developmental blueprint. It's an open-source framework with a strong community and many active contributors.

The inception of LangChain was driven by the aspiration to streamline the development process, making it less daunting for developers to work with LLMs. In fact, this was the main reason I started using Langchain – as an easy means to deploy a proof of concept for a potential client to showcase a chatbot able to chat about a customer's own data.

LangChain is a framework tailored for crafting applications driven by language models. It accentuates context-awareness and reasoning by interfacing with language models and application-specific data. It offers modular components and off-the-shelf chains for ease of use and customization, facilitating a structured assembly of components for higher-level tasks. The framework also encapsulates various modules such as model I/O, retrieval, chains, agents, memory, and callbacks, catering to different aspects of interaction and data handling within your LLM application.

LangChain libraries

The LangChain framework supports two different scripting languages: Python and JavaScript (TypeScript). The essence of LangChain lies in its emphasis on composition and modularity, housing a myriad of individual components. These components can either be amalgamated or utilized independently, showcasing the framework's flexible nature.

Both libraries are easy to use. While the Python LangChain library is more mature and feature-rich, the `TypeScript` library is newer but being developed to closely mirror the Python version's capabilities. Both are designed to be user-friendly, with a focus on simplifying the integration and use of LLMs in various applications.

With such a strong community, there is a huge number of contributors in both libraries, with modules for everything from document loaders to agent blueprints to memory providers as well as integrations from tech providers, large and small.

But to make the best use of LangChain, it's good to start with its basics. In the next section, we'll look at the core components of LangChain.

Core components of LangChain

LangChain offers a variety of modules that can be used to create language model applications. These modules can be used individually in simple applications or combined to create more complex ones. Flexibility is the key here!

The most common components of any LangChain chain are the following:

- **LLM model**: LangChain's core reasoning engine is your language model. LangChain makes it easy to use many different types of LLMs.

- **Prompt templates**: These provide instructions to the LLM, controlling its output. Understanding how to construct prompts and different prompting strategies is crucial, so it's good that we covered this in the previous chapter.

- **Output parsers**: These convert the raw response from the LLM into a more usable format, simplifying downstream processing.

Let's look at each of these core components in more detail.

Working with LLMs in LangChain

There are two different types of LLM models in LangChain, which are called LLMs and ChatModels. These models are used to interface with the different LLM models available. LangChain supports many different types of LLMs and so is a great enabler for trying different types of LLMs and comparing results.

- **LLMs**: This is a language model that takes a string as input and returns a string

- **ChatModel**: This is a language model that takes a list of messages as input and returns a message

LLMs

LLMs in LangChain refer to the pure text completion models that they interface with – for example, GPT-4. These take a simple input as a string and output a string.

Chat models

In comparison, Chat models are tuned for conversations. The LLM APIs use a different interface than pure text completion models. Instead of a single string, they take a list of chat messages as input, which are labeled with a specific role: system, assistant, and user. You've already been introduced to the concept of working with a list of chat messages in *Chapter 3*, in the *Learning to use the ChatGPT API* section. The ChatGPT completions endpoint expects this input.

As we're concentrating on using ChatGPT and creating conversational experiences using LangChain, we'll focus more on Chat Models for the rest of this chapter. If you are interested in carrying out generative tasks, then using the LLM model may be a better fit.

In LangChain, the chat model supports a list of ChatMessages, which have two properties:

- `content`: The content of the message
- `role`: The role of the entity from where the `ChatMessage` originated

The supported roles are as follows:

- `AIMessage`
- `HumanMessage`
- `SystemMessage`
- `FunctionMessage`

You can also create a role manually in the `ChatMessage` class.

The ChatModel provides a `predict_messages()` method to interact with the mode using a list of messages. This would look like the following:

```
from langchain.schema import HumanMessage
text = " Im looking to book a direct flight from New York to London
departing on December 10th and returning on January 5th. Can you
provide me with the available options, including airlines, flight
times, and prices?"
messages = [HumanMessage(content=text)]
chat_model.predict_messages(messages)
```

This should output something that looks like the following:

```
AIMessage(content='Certainly! Here are some available options for
direct flights....')
```

Let's now look at using prompt templates with the LLM.

Prompt templates

As we've seen, prompts can get detailed and long pretty quickly. We also often want to reuse our prompts. This is where the LangChain prompt templates can help a lot. They serve as structured blueprints to craft prompts for all the different tasks we want to carry out.

Prompt templates can be combined, and LangChain also provides prompts for common tasks such as question answering and summarization.

In essence, instead of directly manipulating the prompt string with instructions, we can just pass in the variables for a specific prompt and let LangChain handle the rest. We can also use LangChain prompts to work with our Chat Model prompting, so let's take a look.

The `PromptTemplate` class in LangChain acts as a mold for string prompts, which can be filled with user-defined parameters.

Let's look at an example. You are looking to build an application that analyzes your chat transcripts so you can understand the sentiment of your conversations across the different departments.

We import the `PromptTemplate`, then use the `from_template` function to create the template and corresponding variable placeholders:

```
from langchain.prompts import PromptTemplate
prompt_template = PromptTemplate.from_template(
    "Look at the following conversation {conversation} from the
following service area {service_area} on {event_date_time} and returna
sentiment"
)
conversation = """Customer: My new bike is missing a wheel!\
Chatbot: I'm sorry to hear that. Could I have your order number,
please? ….
"""
conversation_analysis = prompt_template.format(
    conversation=conversation,
    service_area="complaints",
    event_date_time="2023-10-19 14:30:00"
)
```

If you call the `format` method on your template and output `conversation_analysis`, you should have something that looks like the following:

```
"Look at the following conversation Customer: My new bike is missing a
wheel!Chatbot: I'm sorry to hear that. Could I have your order number,
please?... from the following service area complaints on 2023-10-19
14:30:00 and return a sentiment"
```

Notice that you have a completed prompt that is composed of the template and the variables that were passed in.

Chat prompt template

The ChatPromptTemplate works with chat models and enables the creation of conversational prompts, encapsulating a sequence of chat messages with designated roles such as ''system'', 'human', and 'AI'.

Let's take a look at an example. We're creating a health clinic chatbot to assist with general health-related inquiries. We import ChatPromptTemplate, and then create the template by calling the from_messages function:

```
from langchain.prompts import ChatPromptTemplate
chat_template = ChatPromptTemplate.from_messages(
    [
    ("system", "You are a health advisory bot for HealthHub Clinic.
You can answer questions from the patient called {name}"),
    ("ai", "Hi, {name} please ask me your question."),
    ("human", "{user_input}"),
    ]
)
```

You can then use this template as follows; this will create our message list populated with the variables we've assigned:

```
messages = chat_template.format_messages(name="Lucy",
    user_input="What are the symptoms of the flu?")
```

These messages are now correctly formatted as follows:

```
[SystemMessage(content='You are a health advisory bot for HealthHub
Clinic. You can answer questions from the patient called Lucy'),
AIMessage(content='Hi, Lucy please ask me your question.'),
HumanMessage(content='What are the symptoms of the flu?')]
```

You can now use these in a call to your Chat model by passing in your formatted prompts list:

```
chat_model= ChatOpenAI()
Chat_model(messages)
```

There are also out-of-the-box prompts such as a prompt template that uses few-shot examples. The main components of the few-shot template are the examples, which are a list of examples to pass into the prompt and the example prompt itself. The prompt templates give you a lot of options with types of prompts; however, sometimes you may want to create your own prompt template. LangChain also makes this easy with custom prompt templates. Let's look at these in the next section.

Custom templates

Custom prompt templates in LangChain offer flexibility by enabling users to craft bespoke prompts for unique tasks beyond the scope of standard templates. These custom templates become useful when a task demands a distinct prompt format or particular details to steer your language model in the intended direction. Let's look at an example:

```
Consider a scenario where you want to generate responses for a
learning productivity assistant in a chat interface. This assistant
might need to consider the specific task the user is asking about, the
time available, and any particular user preferences or constraints.

Here's an example of what the custom prompt template for this
productivity assistant could look like:
from langchain.prompts import StringPromptTemplate
from pydantic import BaseModel, validator
class ProductivityAdvicePromptTemplate(
    StringPromptTemplate, BaseModel
):
    @validator("input_variables")
    def validate_input_variables(cls, v):
        required_vars = {"task", "time_available",
            "user_preferences"}
        if not required_vars.issubset(v):
            raise ValueError(
                f"Input variables must include: {required_vars}")
        return v

    def format(self, **kwargs) -> str:
        prompt = f"""
        As a virtual productivity assistant, provide advice on:
        Task: {kwargs['task']}
        Time Available: {kwargs['time_available']}
        User Preferences: {kwargs['user_preferences']}

        Advice:
        """
        return prompt
```

Example usage of this template would look something like the following, by passing in the input variables when defining the prompt and calling the `format` function with your required variables:

```
productivity_advisor = ProductivityAdvicePromptTemplate(
    input_variables=[
        "task", "time_available", "user_preferences"])
# Generate a prompt for the virtual assistant
```

```
prompt = productivity_advisor.format(
task="write a research paper",
time_available="2 hours",
user_preferences="focus on quality over quantity")
```

You now have a productivity advice-prompting template that you can reuse and share if necessary.

Using output parsers

Now that we understand an LLM and our prompts, the final core component of a LangChain app is the output parser.

Oftentimes, once we have extracted our response from our LLM, we have specific requirements for a certain format. Our example use case is based on some of the tasks we carried out in *Chapter 2*, in the section named *Creating utterances*. We can use our LLM to create intents and utterances. In these examples, although it appears to respond with JSON, it's actually a string. What would be more useful would be to parse the LLM output string into a Python dictionary.

First, we import all the required libraries and then create our `ResponseSchema` by providing a name and description. We can have as many of these as we require to match our output variables, which we then use to create our output parser by calling the `StructuredOutputParser. from_response_schemas` function.

We can then create our instructions to include in our prompt template by calling `get_format_ instructions()` on the output parser. The instructions are then used in our prompt template. Remember to pass our intent format instructions into our prompt. Let's look at how we would do this in the following code example:

```
intent_schema = ResponseSchema(name="intents",
    description="Format the output as JSON object consisting of a key:
intents, the intents key will be a list of objects with the following
keys: intent, utterance, category")
response_schemas = [intent_schema]
intent_output_parser = StructuredOutputParser.from_response_schemas(
    response_schemas)
intent_format_instructions = \
    intent_output_parser.get_format_instructions()
print(intent_format_instructions)
intent_template = """
Create intents and utterances for a chatbot which will answer
questions about a college, \
Create 10 examples of intents for the following categories: facilities
and course_information \
ensure that each intent has 10 utterances, create 5 long tail and 5
more common utterances \
{intent_format_instructions}
```

```
"""
prompt_template = ChatPromptTemplate.from_template(intent_template)
messages = prompt_template.format_messages(
    intent_format_instructions=intent_format_instructions)
#messages = prompt_template.format_messages(intent_examples=intents)
chat = ChatOpenAI(temperature=0.0, model="gpt-4")
response = chat(messages)
output_dict = intent_output_parser.parse(response.content)
output_dict.get('intents')
```

Once we get the response back from the LLM, we can simply pass our response into our parse function to format our result – `output_dict = intent_output_parser.parse(response.content)` – which will now be a dictionary that we can interact with as normal.

There are a lot of benefits of using the output parser here as it manages any number of specific instructions to be used in the prompt, and the resulting response can be parsed as needed without having to manually format each one. The output parser is the final core component of a LangChain app.

In the next section, we'll look at the concept of chains and the language used to manipulate them.

Understanding LangChain Expression Language

We've covered the core LangChain components. Now, we turn our focus to a pivotal concept that can be used in the creation and execution of these components: **LangChain Expression Language** (LCEL). Let's delve deeper into what LCEL is and understand its significance in the LangChain framework.

What is LCEL?

The concept of chains is that they are the key building blocks of LangChain. A chain is a collection of blocks of functionality and in 'its simplest form, a chain would be made up of a prompt, a call to an LLM, and the processing of the result. We've seen these functional blocks in the previous section. At their most complex, chains can be made up of hundreds of steps.

LCEL provides a declarative syntax and orchestration framework that was created so users could easily create chains of functionality with a standardized way of defining and invocation to make it as easy as possible to run as well as create custom chains.

Key components of LCEL

LCEL offers a lot of functionality out of the box and is based on piping and modularity along with several specific features and a runnable protocol for each component, with each object providing a common interface so you can interact with each one in the same way.

The following is a list of the key features of LCEL:

- **Streaming support**: LCEL enhances streaming efficiency by reducing the delay before the first output is produced. It streams tokens from an LLM to an output parser for quick, step-by-step parsing in real time.

- **Asynchronous and synchronous support**: Chains built with LCEL can be called using both synchronous (e.g., in Jupyter notebooks for prototyping) and asynchronous (e.g., in a LangServe server) APIs.

- **Optimized parallel execution**: LCEL efficiently handles tasks by executing parallel steps in a sequence, such as retrieving documents simultaneously from various sources to minimize delay.

- **Retries and fallbacks**: To make sure your applications are reliable; you can configure retries and fallbacks in any part of the LCEL chain.

- **Input and output schemas**: To help with input and output validation, documentation, and error handling, LCEL chains include Pydantic and JSON Schema, which can be used for validation.

- **Observability**: Out of the box, using LCEL logs the input and output of each step and details of the sequence of steps in the chain, which can then be viewed with the LangSmith server, which we'll cover in later sections.

- **Integration** : For complex chains, accessing the results for each step is often crucial for feedback or debugging. LCEL provides easy integration to LangSmith for monitoring and debugging.

Runnable protocol

LCEL's Runnable protocol is important for custom chain creation. LCEL employs a "Runnable" protocol to streamline the development of custom chains.

The Runnable protocol in LCEL is designed to make it easier to build and use custom processing chains. In simple terms, it's like a set of rules or guidelines that helps in organizing how these chains are created and run. This protocol includes specific methods for tasks such as streaming data and processing in batches, along with their advanced versions for more complex operations. The main purposes of this protocol are as follows:

- **Standardization**: It creates a uniform way to define and use different parts of a system. This means that no matter what component you're dealing with (such as a model or a tool), you use the same set of rules to work with it.

- **Flexibility**: By having a standard approach, it becomes easier to mix and match different components. You can plug different parts into your chain, and they'll work together smoothly.

- **Clarity**: It provides clear guidelines on what kind of data goes into and comes out of each component. This helps in understanding and inspecting how each part of the system functions.

- **Efficiency**: It streamlines the process of building and modifying chains, saving time, and reducing the likelihood of errors.

So, in essence, the Runnable protocol is like a common language or blueprint that helps in building, modifying, and understanding custom chains in LCEL effectively and efficiently.

LCEL syntax

With LCEL, we create our chains differently using the **pipe operator** (|) instead of **Chain**

```
Chain = x | y
```

When LangChain encounters the pipe operator between two objects, it tries to feed x into y.

A simple example of LCEL

To help you see the difference between using LCEL and the traditional approach, let's compare both methods of creating a simple chain.

These two examples achieve the same outcome: setting up a chain with a language model and a default prompt. Let's explore each one:

1. Use the LLMChain class as follows. In this approach, you're explicitly creating an instance of the LLMChain class:

    ```
    llm = ChatOpenAI(()
    second_prompt = ChatPromptTemplate.from_template("tell me a
    joke"))
    chain = LLMChain(llm=llm, prompt=default_prompt)
    ```

    This involves passing in the language model (llm) and the default prompt (default_prompt) as arguments.

2. Create the chain with LCEL by connecting each step with the pipeline operator (|):

    ```
    chain = llm | default_prompt
    ```

    This approach uses the pipeline operator (|) to chain together the language model (llm) and the default prompt (default_prompt). The pipeline operator effectively "'pipes"' the output of one component (the LLM) into the next component (the default prompt processor).

Key differences and benefits of LCEL

There are some key differences and benefits to point out here:

*   **Syntax, style, and readability**: The most apparent difference is in the syntax. The first approach is more verbose and explicit, while the second is more concise and might be more intuitive for those familiar with functional programming or command-line operations where such piping is common. For someone new to LangChain, the explicit LLMChain instantiation might be more readable and easier to understand. The pipeline operator, while concise, might require a bit of getting used to.

- **Extensibility**: Both methods are extensible, but the way you extend them differs. With the LLMChain class, you might extend functionality by subclassing or modifying the class. With the pipeline operator, you extend by adding more components to the pipeline.

- **Underlying mechanics**: While both achieve a similar end result, the underlying mechanics might differ slightly.

- **Advanced features**: LCEL in both Python and Typescript offers benefits such as optimized parallel execution and support for handling errors with retries and fallbacks. LCEL also offers the ability to output the results of separate parts of your chain before the final chain completion, which is great for debugging and sending updates to the end client – for example, if you want to send your chat interface updates during a long execution.

> **Tip**
> Initially, the LLMChain interface was the standard method. However, the more modern method is LCEL. LLMChain is still valid but for new application development, it's advisable to use LCEL for composing chains.

In the following examples in this chapter and the remainder of the book, we'll try and use LCEL wherever possible. In the next section, we'll dive into creating different types of LangChain chains.

Creating different LangChain chains

In this section, we'll explore how to create some of the various chains you can create in LangChain. Our aim is to guide you from basic chains to more advanced chains.

We'll start with an example to introduce the basics of a simple chain. This will provide a foundation for understanding how chains work in LangChain. Next, we'll learn how to create sequential chains, parallel chains, and finally, routing chains.

By the end of this section, you'll have a clear understanding of how to create and utilize different LangChain chains and be equipped with the skills to apply these in your future LangChain projects. Let's dive in.

Basic chain example

In this basic example, we'll show the simplest chain with a prompt and LLM component, which forms the basis of many more complex implementations. In this example, we are using our model to translate text into any language:

```
prompt = ChatPromptTemplate.from_template("""
    Translate this text: {text} to {language}
""")
language = "Spanish"
```

```
text = "what is the capital of england"
runnable = prompt | ChatOpenAI() | StrOutputParser()
runnable.invoke({"text":text,"language":language})
```

Walking through the code, you can see this is a simple chain. We create a `ChatPromptTemplate` instance from a template string. The template string defines the prompt that will be presented to the language model. The prompt asks the model to translate text into a specified language:

```
prompt = ChatPromptTemplate.from_template("""
    Translate this text: {text} to {language}
""")
```

The prompt template is a string that contains the following placeholders:

- `{text}`: This placeholder will be replaced by the text to be translated
- `{language}`: This placeholder will be replaced by the target language

We can then create the runnable, our chain of components that all implement the runnable interface and are designed to be chained together, which is the core concept of LCEL. This line of code creates the `runnable` chain using the `|` operator:

```
runnable = prompt | ChatOpenAI() | StrOutputParser()
```

The `runnable` is a chain of three components:

- `prompt`: This component applies the chat prompt template to the input data
- `ChatOpenAI`: This is the model that sends the prompted text to the `ChatOpenAI` chat models
- `StrOutputParser`: This model parses the output of the `ChatOpenAI` chat models into a string, facilitating easy integration into various applications.

It's likely that you may encounter chains where the `LLMChain` class is instantiated. This is the equivalent of our runnable in this example:

```
llm_chain =LLMChain(llm=llm,
    prompt=PromptTemplate.from_template(prompt_template))
```

The `LLMChain` is now a legacy class, so it's best practice to use LCEL.

In the next section, we'll look at creating a simple sequential chain for a more complex use case that's specific to conversational AI.

Creating a sequential chain to investigate conversational data

Let's start by looking at a simple sequential chain. Sequential chains run a sequence of chains one after another and work well when you have expected input and outputs for carrying out downstream processing.

For this example, imagine we have an electric car subscription company called GreenRide, which uses live chat to handle customer queries on its website. It is interested in creating a chatbot to automate some of its customer conversations. It has a lot of uncategorized chat transcripts. It wants to categorize these chat transcripts into **frequently asked questions (FAQs)** and transactional inquiries so it can understand which queries are complex and which ones are simple. It is interested in using an intent-based **Natural Language Understanding (NLU)** platform for handling transactional inquiries and an LLM to handle the FAQs. With the transactional inquiries, it wants to cluster the conversations by intent so that it can look at each use case in more detail and look at which teams need to be involved.

You have been tasked with looking at this transcript data and carrying out these tasks. So, let's break this down:

1.   Categorize each conversation and split it into two lists of FAQs and transactional conversations.

2.   Look at the transactional list of conversations and cluster these by intent.

To achieve this, we'll use LangChain to process chat transcripts through a sequential chain of operations. These processes are highlighted in the following diagram:

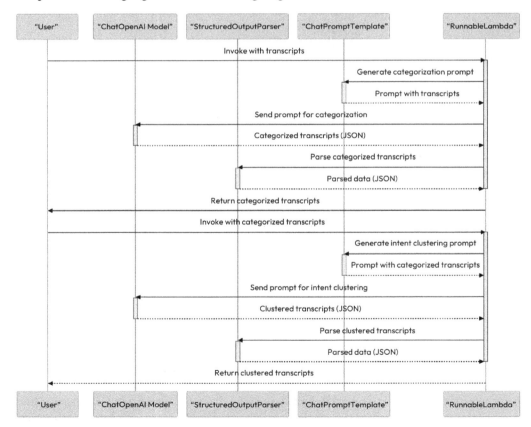

Figure 5.1 – Processes of a sequential chain to investigate conversational data

You will need two chains to carry out the categorization and clustering. The first chain, which uses an LLM and a prompt, will take in the list of transcripts and return two lists of conversations. The second chain will also be made up of an LLM and a prompt and will take in a list of transactional conversations and cluster these by intent type.

In this example, we'll be using LCEL and passing the results of one chain into another for further processing:

1. Start by importing the necessary modules from `langchain` and initialize our `ChatOpenAI` model with temperature `0`, as this isn't going to be a creative task:

    ```
    import json
    from langchain.output_parsers import (
        ResponseSchema, StructuredOutputParser)
    from langchain.prompts import (
        SystemMessagePromptTemplate,
        HumanMessagePromptTemplate, ChatPromptTemplate)
    from langchain.chat_models import ChatOpenAI
    chat_model = ChatOpenAI(
        temperature=0, model="gpt-3.5-turbo-1106")
    ```

2. Chat transcripts are loaded from a JSON file, which will be the data source for processing:

    ```
    with open('transcripts.json', 'r') as file:
        transcripts = json.load(file)
    ```

3. Set up our output schemas to categorize the two conversations into `transactional_transcripts` and `faq`:

    ```
    response_schemas = [
        ResponseSchema(name="transactional_transcripts",
    description="Format the output as JSON list of conversations
    with the same JSON format as the input,add a category key to
    each conversation", type="list"),
        ResponseSchema(name="faq", description="Format the output
    as JSON list of conversations with the same JSON format as the
    input, add a category key to each conversation ", type="list"),
    ]
    ```

    You'll notice that although you have a response schema for your two categorized lists of transcripts, you are only going to be using one, `transactional_transcripts`, for this processing task.

4. Set up your prompt template for categorizing transcripts, including placeholders for dynamic content:

    ```
    transcript_template = "Look at the following chat transcripts
    {transcripts} and categorize them into FAQ and transactional in
    the following format {format_instructions}"
    ```

5. Configure an output parser to structure the language model's output and set up a `ChatPromptTemplate` for the categorization task:

```
output_parser = StructuredOutputParser.from_response_schemas(
    response_schemas)
format_instructions = output_parser.get_format_instructions()
prompt = ChatPromptTemplate(
    messages=[transactional_categorization_prompt_template],
    input_variables=["transcripts"],
    partial_variables={"format_instructions":
        format_instructions},
)
```

6. Create your transactional categorization processing chain combining the prompt, ChatGPT model, and output parser for categorizing transcripts. This is composed using LangChain's chaining mechanism, where each component is linked using the pipe operator (|):

```
chain_one = prompt | chat_model | output_parser
```

This is the first chain complete, so now you can move on to creating the next chain.

7. For the second intent clustering phase, you create new response schemas and an intent clustering prompt template:

```
intent_response_schemas = [
    ResponseSchema(name="transactional_intents",
description="Format the output as JSON list of conversations,
add an intent key to each conversation", type="list"),
]
intent_transcript_template = "Look at the following chat
transcripts {transactional_transcripts} Cluster these
conversations by intent {intent_format_instructions}"
```

8. Set up the second processing chain for intent clustering first, by creating a new prompt and the intent output parser. We can use the same chat model as we used for the first chain although depending on your task, these could be different. The beauty of LangChain is that all the components are interchangeable.

```
intent_clustering_prompt_template = \
    HumanMessagePromptTemplate.from_template(
        intent_transcript_template)
intent_output_parser = \
    StructuredOutputParser.from_response_schemas(
        intent_response_schemas)
```

```
intent_format_instructions = \
    intent_output_parser.get_format_instructions()
prompt_two = ChatPromptTemplate(
    messages=[intent_clustering_prompt_template],
    input_variables=["transactional_transcripts"],
    partial_variables={
        "intent_format_instructions":
            intent_format_instructions},
)
```

9.   The last part of the code helps you to create the second chain with the elements you've created. The output of the transcripts from `chain_one` is passed in:

```
chain2 = (
    {"transactional_transcripts": chain_one}
    | prompt_two
    | chat_model
    | intent_output_parser
)
chain_two_result = chain2.invoke({"transcripts": transcripts})
```

The output should be a dictionary of chat transcripts made up of the transcripts list and the intent and category properties. You can access the first one as follows:

```
print(chain_two_result.get('transactional_intents')[0])
```

There are other improvements you can try here. If you are going to be dealing with larger numbers of transcripts, it might be a good idea to loop over chunks of data so you don't hit on token issues.

You could also add further processing by adding another chain element – for example, creating utterances for intents or categorizing intents for specific departments.

Using LangChain, we can also run chains in parallel. In the next section, let's look at an example.

Utilizing parallel chains in LangChain for efficient multi-source information gathering

There may be situations when you want to run chains in parallel as part of a sequence. For example, imagine a scenario where we want to query multiple information sources simultaneously. This capability is particularly useful when a user asks a complex question that requires pulling data from different types of content, such as news articles, scientific papers, and general web information. The following diagram outlines the processes involved in a parallel chain:

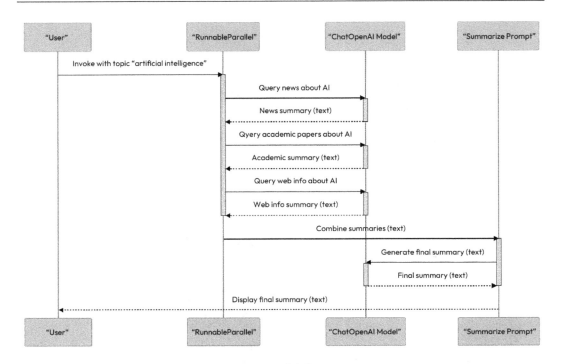

Figure 5.2 – Using a parallel chain in a sequence

The following code example illustrates how you can achieve this approach:

1. Create our model and then create the three separate prompts we're going to use in our chain (news_chain, `academic_chain`, and `web_info_chain`), each using a template to query specific types of content (news, academic papers, and general web information) about a given topic:

```
model = ChatOpenAI(model="gpt-4-1106-preview")
news_prompt = ChatPromptTemplate.from_template(
    "summarize recent news articles about {topic}")
academic_prompt = ChatPromptTemplate.from_template(
    "summarize recent scientific papers about {topic}")
web_info_prompt = ChatPromptTemplate.from_template(
    "provide a general overview of {topic} from web sources")
```

2. Create our runnable parallel chain with the RunnableParallel class, as it is used to run these chains in parallel:

```
parallel = RunnableParallel(
    news = news_prompt | model,
    academic = academic_prompt | model,
```

```
        web_info = web_info_prompt | model
    )
```

3. Invoke the parallel chains with topics (e.g., `"artificial intelligence"`), and each chain will fetch relevant information based on its configuration:

```
results = parallel.invoke({"topic": "artificial intelligence"})
```

This will achieve what you expected. The chains all run in parallel and return their results as a **dictionary** you can then access as follows: `results['web_info'])`.

It may be even more useful to create a summary of all three results to answer your user's question. To achieve this, you need to run the parallel chains and then use these as part of a sequential chain.

4. Create the summarization prompt:

```
summarise_prompt = ChatPromptTemplate.from_template("""
summarize the following information from these different
sources:
News source: {news}
Academic: {academic}
Web: {web_info}
Summary:
""")
```

5. Create our `RunnableSequence` and call our chain with `invoke()`:

```
summarise_chain = parallel | summarise_prompt | model
summarise_output = summarise_chain.invoke(
    {"topic": "artificial intelligence"})
```

In summary, `RunnableParallel` outputs a dictionary with the results of each parallel chain.

We can access those results using the keys we gave the parallel chains (e.g., news, academic, and web info).

We pass that whole output dictionary into the `RunnableSequence`.

Our `summarise_prompt` prompt then has access to those results via the keys to populate its template. So, this allows smooth piping of the outputs of parallel processes into a sequential chain.

Let's move on to a slightly more complex example, where we look intelligently routing our requests to different chains.

Routing chains to answer questions effectively

When creating an LLM-powered application, you may need to call different chains, depending on the type of question or information you need to provide. The `RunnableBranch` allows you to do this.

For an example of this type of implementation, let's go back to our car subscription company: GreenRide. We have an idea of the types of inquiries that we need to handle and these fit into three categories: maintenance, car info, and accounts. However, we'll need to answer these with different chains. Let's look at how this is achieved:

1. First, we import the necessary classes from the `langchain` library, then initialize our model:

```
from langchain.chat_models import ChatOpenAI...
llm = ChatOpenAI(temperature=0, model="gpt-4-1106-preview")
```

2. Next, we need to create the prompts for each department. We create templates for the different types of queries (maintenance, car info, and accounts). Each template is designed to guide the language model to answer in a specific context:

```
maintenance_template = "You are an expert in car
maintenance     {input}"
maintenance_prompt = PromptTemplate.from_template(
    maintenance_template)
car_info_template = "You are knowledgeable about various car
models    {input}"
car_info_prompt = PromptTemplate.from_template(
    car_info_template)
accounts_template = "You are well-versed in account
management     {input}"
accounts_prompt = PromptTemplate.from_template(
    accounts_template)
```

3. We also need a general template:

```
general_prompt = PromptTemplate.from_template(
    "You are a helpful assistant. Answer the question...
{input}"
)
```

This is a fallback template for queries that don't fit into the specific departments.

4. Next, we create the branching logic using a `RunnableBranch`:

```
prompt_branch = RunnableBranch(
    (lambda x: x["topic"] == "maintenance", maintenance_prompt),
    general_prompt,
)
```

This part defines the logic to decide which template to use based on the topic. It checks the topic of the input and matches it to the corresponding template.

5. We then need to create the topic classifier and classifier chain:

```
class TopicClassifier(BaseModel):
    topic: Literal["maintenance", "car_info", "accounts",
        "general"]
```

The `TopicClassifier` class defines a Pydantic data model that represents the output of the topic classification process. It has a single field, `topic`, which is an enum value indicating the classified topics of the user question.

6. Next, we use the `convert_pydantic_to_openai_function` function to convert the `TopicClassifier` Pydantic model into an OpenAI function that can be used with the `ChatOpenAI` API:

```
classifier_function = \
    convert_pydantic_to_openai_function(TopicClassifier)

parser = PydanticAttrOutputFunctionsParser(
    pydantic_schema=TopicClassifier, attr_name="topic"
)
```

7. We then use the bind method to add the converted `TopicClassifier` function to the `ChatOpenAI` object. This effectively makes the topic classifier function available as a callable function within the `ChatOpenAI` context. The `PydanticAttrOutputFunctionsParser` class is used to parse the output of the topic classifier function, which is a JSON object. The `pydantic_schema` argument specifies the Pydantic model to use for parsing the JSON object, which in this case is the `TopicClassifier` model. The `attr_name` argument specifies the attribute of the Pydantic model to extract the classified topic from, which is the topic attribute.

> **OpenAI functions**
>
> We are using OpenAI function calling here to ensure we receive the correct JSON format and schema. It's worth noting that with the latest versions of OpenAI models (`gpt-4-1106-preview` or `gpt-3.5-turbo-1106`), you can set `response_format` to `{ "type": "json_object" }` to enable JSON mode. When JSON mode is enabled, the model is constrained to only generate strings that parse into valid JSON objects. However, this does not ensure the correct schema.

8. We then bring this together in our chain:

```
classifier_chain = llm | parser
```

When the `classifier_chain` is called with a user question, the question is first passed to the `ChatOpenAI` function for topic classification. The output of the topic classifier function is then parsed by the Pydantic attribute output functions parser, and the extracted classified topic is returned, allowing us to define which template to call.

9. Finally, we create our main chain pipeline, bringing all this together:

```
final_chain = (
    RunnablePassthrough.assign(topic=itemgetter("input")
    | classifier_chain)
    | prompt_branch
    | ChatOpenAI()
    | StrOutputParser()
)
```

The final chain first classifies the input, then selects the appropriate template, processes the input with the language model, and parses the output. Finally, it's easy to invoke our chain:

```
final_chain.invoke( {"input": "How do I update my payment method
for my car subscription?"})
```

You should now have a working example that sets up a chatbot that can intelligently route customer support queries to specific departments such as maintenance, car information, and accounts.

At the moment, we are looking at different prompts. We can extend the `RunnableBranch` to route input not only to different prompts but also to entirely different chains. This allows for more complex and specialized processing, depending on the input's topic or other criteria.

Let's assume that in addition to handling maintenance, car information, and account queries, you want to route some queries to a different chain for special handling. For instance, you might have a chain that handles credit control queries such as payments, which needs to tap into a payments service and user history and therefore has a much more complex chain. This is an easy case of just creating the chain and adding to our branching logic in the `RunnableBranch` configuration:

```
credit_control_template = PromptTemplate.from_template(
    "You are a credit control assistant. Provide immediate and useful
credit control advice for GreenRide customers and allow them to pay
their bills .\n\n{input}"
)
credit_control_chain = llm | access_stripe_details credit_control_
template
prompt_branch = RunnableBranch(
    (lambda x: x["topic"] == "credit_control", credit_control_chain),
    general_prompt,
)
```

This is a pretty simple chain and theoretical example that uses a document loader component for looking up user payment account details. This introduces one of the many different modules available in LangChain that we'll look at in the next chapter.

Summary

In this chapter, we've taken our first look at LangChain and you should have a good understanding of what it is, the problems it solves, and its role in simplifying the development of complex LLM applications.

We've looked at the core components of LangChain, such as the LLM models, prompt templates, and output parsers. You should have a good understanding of the LCEL, its key features, and its role in creating and executing chains within the LangChain framework. Finally, the chapter has covered the various examples of LangChain chains, ranging from basic to complex, to illustrate the practical application of the concepts discussed.

In this chapter, we've only scratched the surface of Langchain, but hopefully, we've built a solid foundation for you to engineer LangChain applications and enable you to move on to the more complex concepts and applications of the LangChain library. In our next chapter, we'll look at other concepts such as memory as well as agents, modules, and other tools, particularly LangSmith.

Further reading

The following links are a curated list of resources to help you with the topics discussed in this chapter:

- `https://www.chatsplitter.com/`
- `https://jupyter.org/`
- `https://docs.pydantic.dev/latest/`
- `https://platform.openai.com/docs/guides/text-generation/function-calling`
- `https://python.langchain.com/docs/get_started/introduction`
- `https://js.langchain.com/docs/get_started/introduction`

Advanced Debugging, Monitoring, and Retrieval with LangChain

In this chapter, we're going to look at some of the more advanced LangChain subjects and their applications. You'll gain an understanding of some important processes, such as debugging LangChain and using the new LangSmith platform. You will gain an understanding of the power of agents and tools and see how they can be used to give your agent superpowers. You'll look at out-of-the-box tools and create a custom tool for an agent. You'll then learn about memory in the context of creating **large language model** (**LLM**)-powered conversational experiences and see how this can be implemented with LangChain.

This chapter aims to build on the previous chapter by looking at more advanced concepts, enabling you to create more complex LangChain applications.

In this chapter, we'll cover the following topics:

- Debugging and monitoring LangChain
- Leveraging LangChain agents
- Exploring LangChain memory

Technical requirements

We will be using ChatGPT extensively in this chapter, so you will need to sign up with a free account. If you haven't created an account, go to https://openai.com/ and click **Get Started** at the top right of the page or go to https://chat.openai.com.

The examples require Python 3.9 and Jupyter Notebook to be installed, both of which you can download from `https://jupyter.org/install`. Jupyter Notebook is also included with Anaconda. You can find this chapter's code in this book's GitHub repository: `https://github.com/PacktPublishing/ChatGPT-for-Conversational-AI-and-Chatbots/chapter6`.

Debugging and monitoring LangChain

By now, you've probably realized that even with the simpler chains, there is still a lot going on with LangChain under the hood. Often, when you're constructing your prompts from different sources, you'll need to see what the inputs and outputs of your language model are to understand and iterate on them.

So, it's important to see what's going on inside your application so that you can build, monitor, and debug your application. LangChain offers several different ways to achieve this.

Let's explore these options to gain a clearer understanding of how they can enhance your application's functionality. We'll start with the simplest options and progress to the most comprehensive option: LangSmith.

Understanding tracing techniques

LangChain offers some simple ways to output the different steps of a chain.

If you're working on the examples that are presented in this book using Jupyter Notebook or by simply running Python scripts, you have a few options to provide tracing output:

- **Verbose**: Setting `verbose` to `True` creates a global setting and will output the steps of a chain at the highest level so that we can see the different LLM calls that are occurring. You can set this as follows:

```
from langchain.globals import set_verbose
set_verbose(True)
```

- **Debug**: Setting `debug` to `True` creates a global setting that outputs all the inputs and outputs generated by components with callback support (chains, agents, models, tools, and retrievers) and gives the most comprehensive information:

```
from langchain.globals import set_debug
set_debug()
```

- **Verbose logs**: You may also want to activate verbose logs on a specific chain as the global output can be too much if you're working with multiple chains. If you're using a chain object, then this can be achieved as follows:

```
conversation_with_summary = ConversationChain(
    llm=llm,
    verbose=True,
)
```

- **Callbacks**: If you're using LCEL to create your chain and you want to set `verbose` to True, then you'll need to use callbacks. In the following code example, we're passing in the config with `ConsoleCallbackHandler` when we call `invoke` on `chain`:

```
from langchain.callbacks.tracers import ConsoleCallbackHandler
from langchain.schema.runnable import RunnableLambda
def fun_1(x: int) -> int:
    return x + 1
def fun_2(x: int) -> int:
    return x * 2
runnable = RunnableLambda(fun_1) | RunnableLambda(fun_2)
runnable.invoke(1, config={'callbacks':
    [ConsoleCallbackHandler()]})
```

The preceding code creates a chain of runnables and then invokes it. We use the | operator to chain two runnables. This means that the output of `fun_1` will be passed as the input to `fun_2`. Here, `runnable.invoke(1, config={'callbacks': [ConsoleCallbackHandler()]})` invokes the chain starting with input 1. The configuration specifies that `ConsoleCallbackHandler` should be used.

The output will show each step of the chain as it's run. Part of the output can be seen here:

```
[chain/start] [1:chain:RunnableSequence] Entering Chain run with
input:
{
  "input": 1
}
[chain/start] [1:chain:RunnableSequence > 2:chain:fun_1]
Entering Chain run with input:
{
  "input": 1
}
```

These techniques are fine if you're still developing using a notebook or Python script. However, if you want to persist your logs, you require more granular information and an intuitive interface, or you're looking to go into production with your application, it's probably time to look at using LangSmith. We'll cover this in the next section.

Introducing LangSmith

LangSmith is currently an early product in beta. Along with being an early release, it is the new tool for working with LangChain and it's likely to be something you'll work with a lot. It simplifies debugging, testing, monitoring, and evaluating your LangChain project. This section will only briefly introduce LangSmith, so you're encouraged to look at the documentation as it is being constantly updated.

Setting up LangSmith

Getting started with LangSmith involves a few straightforward steps:

1. First, you'll need to create a LangSmith account and generate an API key.

    I.   Visit the LangChain website at `smith.langchain.com`. Follow the instructions on the site to sign up and create a new account.

    II.   Once your account has been created, log in to the LangSmith console using your new credentials.

    III. After logging in, navigate to the **Settings** page within the LangSmith console. Here, locate the option to create an API key and generate one.

2. Next, you need to configure your development environment so that they can integrate with LangSmith. This is simply a case of setting the following variables:

    ```
    os.environ['LANGCHAIN_TRACING_V2'] = "true"
    os.environ['LANGCHAIN_ENDPOINT'] = "https://api.smith.langchain.com"
    os.environ['LANGCHAIN_API_KEY'] = "Your Langchain API key"
    ```

Once you have set these up, each run will mean that the respective traces will be logged to your default project. A **run** represents a single unit of work within your application.

It's easy to target a specific project by setting the following command:

```
os.environ[' LANGCHAIN_PROJECT'] = "Your project name"
```

You can then look at each run and drill down into the specifics of each to see exactly what occurred in each step.

Leveraging LangChain agents

LLMs are incredibly powerful, but they also have some fundamental flaws. We've discussed their limitations extensively; remember that they only have knowledge up to when their training data was cut off for ChatGPT, which was in late 2021. Surprisingly, technology this advanced is also not very good at other computational tasks, such as basic math, logic, or even the ability to look up other information that they don't know about.

To compensate for these drawbacks, we need another technique or method to help out where our LLM needs assistance. These solutions are called **agents,** and, in this section, we'll look at what they are, how they work, and how to use them with LangChain.

What is an agent?

Agents provide tooling for LLMs to carry out tasks that they can't carry out themselves effectively. There are several different types of agents that are designed to work with different models, such as OpenAI LLMs, and provide support for different tasks and use cases, such as XML or JSON handling.

Each type has specific use cases and it's worth looking at the LangChain information about each agent so that you can decide which one is best for your use case.

To create an agent, we need to consider three different components:

- LLM
- A tool to carry out a task
- An agent to control the task

The LLM will handle understanding and generating responses, the specialized tool will execute specific tasks, and the agent will orchestrate these tasks.

What are LangChain tools?

LangChain tools allow agents to carry out their defined tasks by interacting with external systems. A tool consists of a schema for the tool with information about what inputs the tool expects and a function to call to carry out the task. LangChain provides lots of built-in tools and lets you create custom tools, something you'll likely want to do at some point. Along with individual tools, LangChain also provides the concept of toolkits, which are groups of tools you can use to carry out specific tasks.

Let's look at what makes up a tool:

- A name and description to document its intended usage
- A JSON schema to define input parameters
- A run function that gets called with inputs
- A `return_direct` flag indicating if outputs should go to the user

It's important to include as much of this information as possible and keep tools simple so that they're easy for the LLM to use. The name, description, and schema will all be used in the prompt to the LLM so that it knows what action to take. These need to be clear and concise.

An introduction to OpenAI tool calling

Before we dive into the next example, it's important to cover OpenAI functions/tools so that we understand why they exist and how they can be used. Just to confuse matters, these used to be called **functions** but now they are called **tools**, which makes more sense.

Before function calling was released, you would ask ChatGPT a question such as *"What's the latest news in London?"* and you wouldn't get a useful response. As we know, it's no good at returning up-to-date information because it's not connected to the internet and can't pull or scrape data.

We could write some code to go and call a news service, but we'd need to pass the location and any other parameters and go back to our conversation. This isn't easy to do, especially if we want to use parameters from the conversation to call our service.

Function calling was created and is supported by the ChatGPT Turbo and GPT4 models to overcome these problems and provide an easy way to answer questions by easily calling external APIs and extracting structured data. These models are trained to look at the functions and intelligently decide when to call them. The basic process to implement functions is as follows:

1. Trigger the model with a user query and predefined functions. The model may call functions, generating a JSON object with possible imaginary parameters.

2. Decode the JSON in your code and execute the function with any given arguments.

3. Input the function's response back to the model so that it can be summarized for the user.

Let's look at a simple example.

Looking at a simple OpenAI tool example

We pass our function definitions to our chat completion via the `tools` parameter on a chat completion call, which accepts an array of tools represented in standard **JSON** format. The following is a definition for a tool that ChatGPT will hopefully consider and call to get the latest news articles about a specific topic:

```
tools = [{
    "type": "function",
    "function": {
      "name": "get_news_articles",
      "description": "Retrieve news articles based on a specific
topic and date range",
      "parameters": {
        "type": "object",
        "properties": {
          "topic": {
            "type": "string",
```

```
                "description": "The topic or subject of interest for the
news articles"
                },
            "required": ["topic"]
        }
      }
    }
  ]
```

We provide the name of the tool, which will force our function to be called by the model and provide as detailed a description as possible. Remember that this description will be used in the context of the prompt, so the more accurate and detailed it is, the better. When it comes to tools, try and describe what your function is going to do as this is how the model has been trained. Then, we add the description for the parameters that we are going to send to the function. In our case, we are passing a `topic` string.

Now, when we try a chat completion by asking our ChatGPT model about the latest news in London, the model will know to call our function with the specific parameters we've defined in the tool definition. If we had more parameters defined in our tool, ChatGPT would ask clarification questions until it has all the information it requires.

The tools are passed into our completion call as follows:

```
response = client.chat.completions.create(
        model="gpt-3.5-turbo-0125",
        messages=messages,
        tools=tools,
        tool_choice="auto",
    )
```

Notice that we have set `tool_choice` to `auto`, which tells ChatGPT to decide whether to use a tool.

This should output something like the following, showing `ChatCompletionMessage` with no content specified. However, we should have `ChatCompletionMessageToolCall` in our tool calls for our function. We also have the `topic` argument and our function's name:

```
ChatCompletionMessage(content=None, role='assistant',
    function_call=None,
    tool_calls=[
        ChatCompletionMessageToolCall(
            id='call_ewHp8R8GUAHY6Dy4UXdEej3B',
            function=Function(
                arguments='{"topic":"London"}',
                name='get_news_articles'),
            type='function')])
```

Here, we can see that ChatGPT has decided to call our function; we can do some tests if it decides otherwise. If we change our question to something like *"Explain quantum physics,"* then we should get `ChatCompletionMessage` with content answering the question:

```
ChatCompletionMessage(content="Chess is a classic strategy board game
played ....
```

As we can see, ChatGPT decides on the tools to be used to satisfy this question and lists these in our `tool_calls` list. We can cycle through this and call each function with arguments if ChatGPT has extracted these from our question. After, we can add the results to our messages list and let ChatGPT summarize the results for the user.

Don't forget that you can also set `tool_choice={"type": "function", "function": {"name": " another_response_function"}}` in our chat completion call to force ChatGPT to use a specific function instead of letting it decide for itself.

The other powerful application of functions/tools is to make it easy to get ChatGPT to create deterministic, perfectly formed data. In *Chapter 2*, we were prompting for user utterances or datasets in JSON. We can now use tools to create perfect output by creating a function definition to do this for us. I recommend that you take a look at some of the training data creation tasks in *Chapter 2* and create a tool to achieve the same outcome.

A lot is going on with OpenAI functions, but this should at least give you enough understanding of the benefits and steps involved in using them. Now, let's look at LangChain tools and see where LangChain streamlines the use of OpenAI functions.

Plug-and-play LangChain tools for immediate integration

LangChain provides a lot of out-of-the-box tools, which makes it easy to provide tools for your agent. You can find these listed at `https://python.langchain.com/docs/integrations/tools/` in the docs, which are created and maintained by vendors and community members. They cover a wide range of functionalities, from search engines to ways to interact with filesystems and ways to interact with big cloud services vendors such as Google Drive.

An out-of-the-box tool example

Let's look at one of my favorite tools: the Tavily Search API. Tavily Search is a search engine that's optimized for LLMs to provide the best search results. So, it's a good example of an LLM tool and one that you're likely to use. While looking at how to create this tool, it's also useful to look at its details so that we can understand what components make for an effective tool. First, we need to install the required libraries:

```
pip install -U langchain-community tavily-python
```

Next, import the library and create the API keys that we'll need:

```
os.environ['OPENAI_API_KEY'] = OPENAI_KEY
os.environ["LANGCHAIN_API_KEY"] = "TAVILY_API_KEY "
```

Now, we can create the tool by creating an instance of the `TavilySearchAPIWrapper` class, which acts as a wrapper around the Tavily Search API. We can pass this instance into our `TavilySearchResults` class:

```
tavily_search = TavilySearchAPIWrapper()
tavily_tool = TavilySearchResults(api_wrapper=tavily_search)
```

You can inspect the different properties of the tool by printing them out via `print(tavily_tool.name)`, `print(tavily_tool.description)`, `print(tavily_tool.args)`, and most importantly carry out a search by calling `invoke()`:

```
tavily_tool.invoke({"query": "how many named storms have there been
this year in the UK"})
```

When you run this, you should get a response that contains up-to-date information on the search. You should see an output similar to the following:

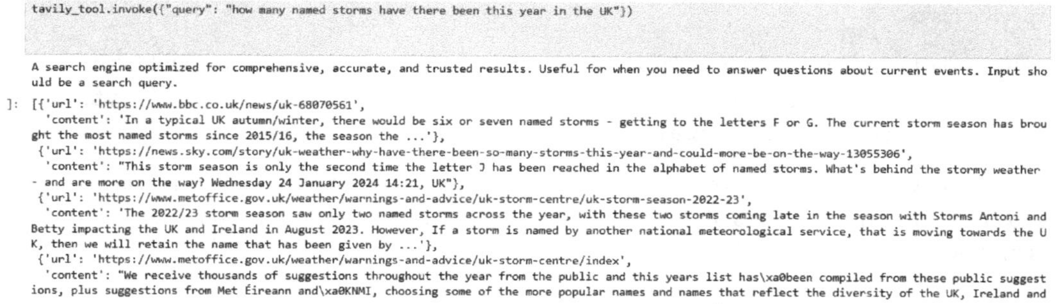

Figure 6.1 – Tavily Search output

In the next section, we'll look at how to use the tool with an agent.

Using our tool with an agent

At this point, we've created a tool, but we aren't using it yet. To do this, we need to create an agent. In our example, we're using an OpenAI Functions agent. We covered OpenAI functions in the previous section; in this section, we'll see how easy it is to implement this with LangChain. The agent can be created using the `create_openai_functions_agent()` function, which takes three arguments:

- `llm`: A language model that supports OpenAI's function calling API
- `tools`: A list of tools that the agent can use

- `prompt`: A prompt that the agent will use

We pass these three parameters to create our agent:

```
open_ai_agent = create_openai_functions_agent(llm, tools,prompt)
```

Now that we have this agent, we'll need to create an `AgentExecuter` class to run it:

```
open_ai_agent_executor = AgentExecutor(agent=agent, tools=tools)
```

We should then be able to call the `invoke()` function and return some results that contain up-to-date information:

```
result = open_ai_agent_executor.invoke({"input": "how many named
storms have there been in 2024 in the UK"})
```

If you print out the result, you should get an answer to the question with up-to-date information showing that our agent and tool are working correctly. Remember to look at the logs in Langsmith so that you can follow what's going on there.

You can go even further with LangChain's plug-and-play tools by using the toolkits provided by the LangChain ecosystem. These toolkits offer collections of tools that are designed to be used together for specific tasks.

Now that we've looked at a prebuilt toolbox and covered agents, you can see how powerful LangChain agents are. You're encouraged to look at some of the other tools and agent types at `https://python.langchain.com/docs/integrations/tools/`.

You'll likely have a requirement that needs a different type of tool. The good news is that it's also easy to create custom tools. Let's dive in and look at a specific example so that you understand the steps involved.

Creating a custom weather tool

For our example, we're going to create a custom tool to get the current weather for a location using Open-Meteo, the free weather API Meto.

Our tool will take two arguments, `latitude` and `longitude`, so that we can look up the weather for a specific location. Both arguments are required.

The first thing to consider is that we need an agent type that supports multiple inputs. So, let's go with `StructuredChatAgent`. Let's jump into the code:

1. First, import the necessary modules. The script imports various Python modules, including those for making HTTP requests, handling data, and specific LangChain components for tool and agent creation:

    ```
    import os
    ```

```
from langchain.tools import BaseTool
from typing import Union
from langchain.agents import (AgentExecutor,
    create_structured_chat_agent)
from langchain import hub
from langchain_openai import ChatOpenAI
```

2. Set up the necessary environment variables for OpenAI and LangChain API, your LangSmith key for logging, as well as other config details.

3. LangChain tools are just functions that we call from our agent. To create one, we must follow the same basic pattern. This involves subclassing `BaseTool`, implementing `run()`, and passing instances of our tools when creating the agent. The custom tool class, `GetWeatherByLocationTool`, extends the `BaseTool` class, providing it with specific functionality to fetch weather information based on latitude and longitude. We must also provide a name and description so that our LLM knows how and when to use the tool:

```
class GetWeatherByLocationTool(BaseTool):
    name = "Weather service"
    description = " A weather tool optimized for comprehensive
up to date weather information. Useful for when you need to
answer questions about the weather.  Use this tool to answer
questions about the weather for a specific location. To use the
tool, you must provide at the following parameters"
 "['latitude', 'longitude']. "

    def _run(self, latitude: Union[int, float], longitude:
Union[int, float]) -> str:
        . . .
            return result_string
        else:
            return "Failed to retrieve weather data."
```

This tool's `_run()` method performs the actual logic to fetch weather information using the Open-Meteo API and returns the results as a string.

4. Now that we've created our tool, we can initialize the language model and the custom tool, and then create the agent using the `create_structured_chat_agent()` function, which combines the language model with the tool for handling specific queries:

```
llm = ChatOpenAI(model="gpt-3.5-turbo-1106")
tools = [GetWeatherByLocationTool()]
prompt = hub.pull("hwchase17/structured-chat-agent")
agent = create_structured_chat_agent(llm, tools, prompt)
```

5.   Create an `AgentExecutor` class, which is responsible for executing the agent with the provided tools. Then, invoke the agent with a specific query, like so:

```
executor = AgentExecutor(agent=agent,
    tools=tools,,handle_parsing_errors=True)
output = executor.invoke({"input": "What's the weather like in
waddesdon"})
print(output)
```

Hopefully, when you've run this, you'll get something like the following:

```
{'input': 'whats the weather like in waddesdon', 'output': 'The
current weather in Waddesdon is 5.6°C with no rain.'}
```

The cool thing is that we're passing in a location and the LLM knows the latitude and longitude coordinates to use for that location and uses these in the call to the custom tool.

With this, you've seen how powerful LangChain agents are when to create a custom tool, as well as rely on some of the powerful out-of-the-box tools. The possibilities are endless!

In the next section, we'll look at a more advanced requirement for your ChatGPT-powered agents, which is to give them memory. We'll learn how to achieve this with LangChain.

Exploring LangChain memory

As we work with LLMs, a key challenge emerges as they cannot inherently recall past interactions. So, in essence, they are stateless. A stateless operation won't persist information from one request to the next, which is a problem if you want to create a chatbot. The way around this is to add the full conversation to the context. The ChatGPT client itself will be passing the full conversation into each prompt as it progresses.

What we want our ChatGPT applications to do is offer stateful interactions where information is remembered across requests and sessions. To achieve this, we need to use a memory mechanism. Different representations of memory will then be included in an LLM prompt. This section is dedicated to exploring the concept of memory in the context of LLMs. We will delve into different types of memory and the challenges you'll face in using memory with LLMs before taking a deep dive into how to use LangChain to provide memory capabilities to ChatGPT-powered applications.

> Tip
> The ChatGPT UI uses memory under the hood, introducing state by passing the message history every time a prompt is provided. This also allows you to select and restart previous conversations.

Exploring the different types of memory applications

If you think about having a conversation – whether this is just chatting about a subject, a more personal conversation, or a conversation to achieve a transactional outcome – it quickly becomes apparent that there could be different types of memory at play. The obvious need for memory is to store what has been said during an active conversation. However, other memory specifics that might be needed are a memory of historical conversations, a memory of specific information related to a user, or knowledge of a specific subject matter related to a user or application. Let's look at the different memory types in the context of a working conversation.

Memory in active conversations

One of the following forms of memory comes into play within an active conversation:

- **Conversational buffer memory**: This memory type archives the entire text of ongoing dialogs, including all inputs and responses. It preserves the full context of the conversation so that it can pass each prompt to the LLM.

- **Conversational buffer window memory**: A more complex version of buffer memory, this maintains only a specified number of the most recent interactions, discarding older ones or combining them as a summary. It functions as a short-term memory, optimizing token usage as a conversation progresses over multiple turns.

Historical memory

Historical memory can be broken down into two types:

- **Past conversations**: This includes long-term memory, which could be previous conversations and sessions for this specific user. This user will likely be known to your agent through authentication. Historical conversations can be as simple as a list of previous interactions, or they may be a more complex implementation where interactions are queried before use.

- **Past conversations summary**: This method builds on historical memory and condenses previous interactions before integrating them into the model's history. It's designed to decrease token use and streamline memory management while preserving essential elements of the conversation.

Personalized memory

Personalized memory may encompass broader information such as user preferences and historical data organized either in structured formats such as knowledge graphs or unstructured formats such as document collections. This type of memory may also provide memory bank mechanisms to fine-tune the memory to provide the best support to the LLM.

Understanding memory challenges

For all these different types of memory, the key resource we must consider is the context window size of the LLM we are using. This will determine the size of the working conversation, including the prompt and result, so each of these types of memory that you choose to implement will need to be considered in the overall size of the context window. As a ChatGPT developer, you need to treat this context window size as a finite resource, so deciding what information should be within this window can become a challenging aspect of creating complex LLM applications.

The good news for the development of your ChatGPT applications is that OpenAI has released increasingly large context windows for their models. For example, the latest GPT-4 model supports more than 32k tokens, which is roughly 24k words.

The bad news is that having bigger, increasingly larger contexts may not guarantee improved performance. The following are some other factors that may influence response accuracy or have other ramifications:

- **Position**: The relative position of the information may influence the results, so strategically placing the most important information at the beginning or end of the context window could improve the model's output

- **Clarity**: A larger context may mean accuracy declines. It's wise not to take the attitude that the bigger the context available, the more you can add to it. Loading up as much information as you can may result in less accurate or relevant responses.

- **Knowledge precedence**: Many LLM implementations will involve domain-specific knowledge. As an LLM developer, you may need to decide whether this knowledge takes precedence over conversational memory when context memory resources are at a premium.

- **Cost**: Despite OpenAI's attempts to reduce costs with each released GPT model, these larger contexts enable you to use more tokens and costs can skyrocket as a result if you're not careful.

> Tip
> When using memory, consider not only selecting the appropriate information but also structuring the information in a way that enhances model performance.

Introducing memory usage techniques

There are several techniques you can use to manage memory that have varying degrees of complexity and ease of implementation, with some looking to streamline active conversational memory and others looking to assist with history and knowledge. Let's look at the different memory usage techniques we can use:

- **Rolling active conversation memory**: The rolling window technique maintains a dynamic window of the most recent interactions or messages, ensuring that the model remains focused on the most current context. The following figure shows the process of this type of memory implementation:

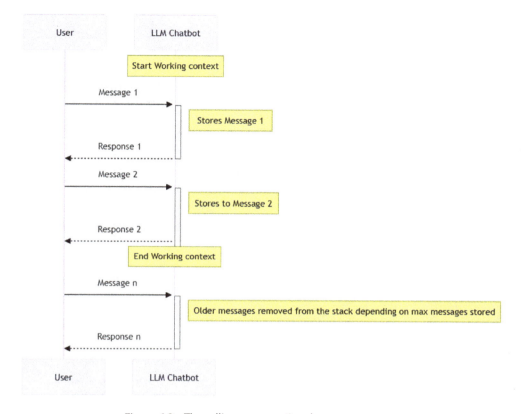

Figure 6.2 – The rolling conversational memory process

As new messages arrive, older data is removed from the window, effectively limiting the model's exposure to past messages. This approach is both straightforward and efficient. However, this is a basic approach, and it carries the risk of discarding valuable message history in the context without having any way of knowing what's just been removed and how this will affect the conversation.

- **Incremental summary of active conversation memory**: This approach extracts the key takeaways from a conversation as a summary and feeds this to the context instead of the entire dialog, along with the most recent messages. This method reduces the risk of data loss and allows for a concise summary while still keeping the theme of the conversation. The following figure shows the incremental memory implementation process:

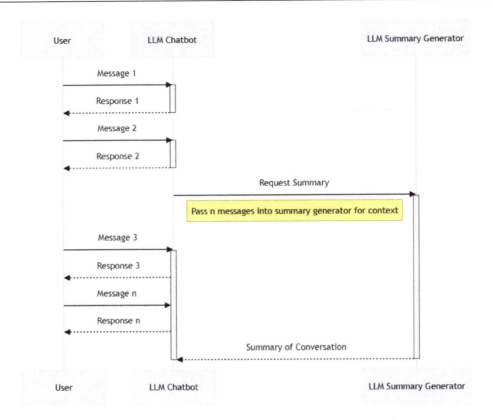

Figure 6.3 – The incremental summary technique

However, If you're summarizing content over multiple conversation calls, this can cause a steady dilution of the nuances of the conversation. Additionally, creating these summaries will need further LLM calls, increasing computational costs

- **Enhancing context with a memory bank:** There's an emerging technique known as a memory bank, which is a memory mechanism tailored for LLMs. This type of implementation will recall multi-turn conversations, personal information, and summarized events from its memory storage mechanism. The following figure outlines the process that's involved in using a memory bank:

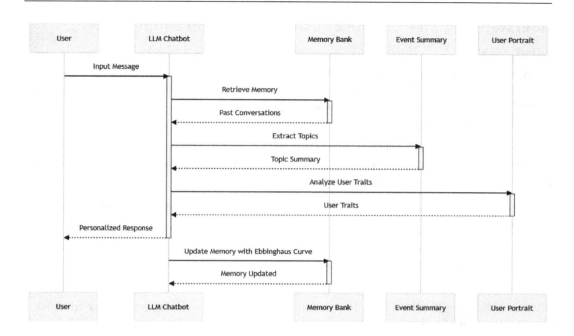

Figure 6.4 – Memory bank technique

The application will manage what's stored and forgotten by using a memory updater inspired by the Ebbinghaus Forgetting Curve theory, all while using its LLM calls to decide what to use in the current conversation, hence continually evolving its context. Bringing all this together should hopefully provide a much more personalized experience. Implementing an application with a memory bank would be a complex undertaking.

In the *Understanding memory challenges* section, we outlined that there may be some issues with larger contexts. It may be that a model with a 4k token limit is just not going to be enough to support everything you need to fit into the context, so it may be a good idea to look at a GPT model with a larger context.

One of the easiest ways to overcome memory constraints is to break more complex tasks into smaller prompts. This technique helps simplify complex prompts and often gets better results.

Hopefully, you now understand memory in terms of creating conversational experiences and some of the available techniques and can appreciate some of the challenges and complexities of adding memory. In the next section, we'll see how LangChain can be used to add memory to an LLM application.

Understanding an example of using memory in LangChain

In its simplest form, a conversational system must have the capacity to directly access a range of past interactions so that they can be used to display conversation history in a chat interface or – in our case – passed to the LLM. In this example, we're going to look at using LangChain memory for a simple conversational interface for a live music assistant. The main aim is to show how we can use memory in a simple conversational application.

LangChain provides several different integrations for supporting memory, from in-memory to persistent databases. In the following example, we'll be using the `ConversationBufferMemory` class, which is the most basic form of memory in langChain.

Fundamentally, our memory system should facilitate two primary functions: reading and writing to memory.

After receiving initial inputs from the user and before initiating its core logic, a chain reads from its memory system so that previous interactions can be used in the prompt.

Once the core logic has been executed but before delivering the response, the outputs and inputs are written to memory by the chain. This action ensures that this information is available for reference in future interactions. In essence, this is similar to the way a conventional conversational AI platform will manage a conversation.

These processes are illustrated in the following example:

1. First, we must import the necessary classes from the `langchain` library and then initialize our model:

    ```
    llm = ChatOpenAI(temperature=0.0,
        openai_api_key=os.getenv("OPENAI_API_KEY"),
        model="gpt-4-1106-preview")
    ```

2. Next, a `ConversationBufferMemory` object must be initialized to manage the conversation history. The `return_messages=True` parameter ensures that the memory returns a list of messages when queried. By default, these are single strings:

    ```
    memory = ConversationBufferMemory(return_messages=True)
    ```

3. Now, we must set up a `ChatPromptTemplate` template that structures the conversation. It includes a system message that defines the role of the bot (a music fan events chatbot), a placeholder for historical messages, and a structure for receiving human input.

    ```
    prompt = ChatPromptTemplate.from_messages([...])
    ```

4. At this point, we can create the chain using LCEL. The chain is composed of a series of operations. The chain combines memory loading, prompt processing, and language model response generation:

```
chain_with_memory = (
    RunnablePassthrough.assign(
        conversation_history=RunnableLambda(
            memory.load_memory_variables) |
            itemgetter("conversation_history")
    )
    | prompt
    | llm
)
```

> **Tip**
>
> Using `RunnablePassthrough` allows you to continue adding a dictionary of data as you move through the steps of a chain sequentially. In this example, we load it into a history key to be used in the prompt.

Here, `RunnablePassthrough.assign(...)` is used to create an extra key, after which the conversation history is assigned to this key. Then, `RunnableLambda(memory.load_memory_variables) | itemgetter("history")` fetches the conversation history.

The chain then applies the prompt template and sends the processed input to the LLM for generating a response.

We can initiate the first interaction by inputting `hi` and waiting for a response. We can then save the input and output in memory for context in future interactions:

```
inputs = {"input": "hi"}
response = chain_with_memory.invoke(inputs)
 memory.save_context(inputs, {"output": response.content})
```

The subsequent interactions continue. This is repeated for each new user input, thus maintaining the context of the conversation:

```
inputs = {"input": "My name is Adrian and I like the high flying
birds"}
response = chain_with_memory.invoke(inputs)
memory.save_context(inputs, {"output": response.content})
inputs = {"input": "I'm looking for gigs in London this summer"}
response = chain_with_memory.invoke(inputs)
memory.save_context(inputs, {"output": response.content})
```

Remember to retrieve the conversation history so that we can load the stored conversation variables for review or further processing:

```
memory.load_memory_variables({})
```

This example was simple as `ConversationBufferMemory` is an in-memory store that stores all the interactions. LangChain offers several different memory integrations. In the next section, we'll look at how these can provide better memory support.

Exploring more advanced memory types

Let's look at some more advanced types of memory available in LangChain.

In-memory stores

Since we're learning to create conversational experiences with ChatGPT to support more complexity, we'll likely need to support multiple-turn chats. With a longer conversation, you can imagine that buffer size might become a concern. It might be wise to look at using `ConversationBufferWindowMemory`, which allows you to store a specific number of interactions by passing the k value on creation:

```
memory_with_5_interactions = ConversationBufferWindowMemory( k=5)
```

This way, you're keeping a sliding window of the 5 latest interactions so that the buffer doesn't get too large.

Having your conversation interactions fixed to a specific number in memory could be limiting, so you could look at using `ConversationSummaryMemory` to help with summarizing the conversation. This type of memory can be instantiated as follows:

```
memory = ConversationSummaryMemory(llm=llm, return_messages=True)
memory.save_context({"input": "hi I'm adrian"},
    {"output": "whats up"})
memory.load_memory_variables({})
```

Here, `ConversationSummaryMemory` collects and updates an overview of the ongoing dialog. This summary, which is stored in memory, can then be used in prompts or chains to provide a quick recap of the conversation so far.

You can take this even further by combining the two concepts of a conversation buffer and conversation summary by using `ConversationSummaryBufferMemory`.

This type of memory stores the latest interactions based on the token length that was set when you instantiated the memory. Instead of just deleting interactions when they are outside of the token limit, the memory stores them as a summary and uses them in subsequent interactions. The memory can be used in `ConversationChain` as follows:

```
conversation_with_summary = ConversationChain(
    llm=llm,
```

```
    # Notice the max_token_limit is very low so you can see the output
of the summary
    memory=ConversationSummaryBufferMemory(
        llm=llm, max_token_limit=150),
    verbose=True,
)
response = conversation_with_summary.predict(input = "My name is
Adrian and I like 1990s music, can you remember my 3 favourite bands,
Nirvana, Offspring and Underworld")
response = conversation_with_summary.predict(input = "What's my second
favourite band")
```

Notice that we set `verbose=True` and `max_token_limit=150` so that we can see a summary and message history in memory. If we output the memory with

`memory.load_memory_variables({}`, we should get something similar to the following:

```
{'history': [SystemMessage(content='Adrian, who likes 1990s ...'),
  HumanMessage(content="What's my second favourite band"),
  AIMessage(content="Your second favorite band is Offspring,
Adrian...")]}
```

Notice that the summary is loaded into the system memory and that there is a list of `HumanMessage` and `AIMessage` object interactions stored in memory.

Using persisted memory

The previous types of LangChain memory implementations we've looked at are great for handling live conversations. However, this memory does not persist between sessions. We touched on this type of approach where you can use historical conversations as part of your context in the *Introducing memory usage techniques* section. So, let's look at how to do this with LangChain.

In this example, we're going to use a vector store and embeddings. Don't worry about the details of these concepts just yet – we'll get into all that in the next chapter when we explore **retrieval-augmented generation (RAG)**. Here, we're using embeddings and vector stores to keep track of past conversation turns so that we can search over them.

We can refer to relevant information from earlier interactions when sending prompts to the LLM to provide conversation history in the context. Let's get started:

1. First, we must install the required packages, most notably ChromaDB, which is the database we'll be using for storing our embeddings:

    ```
    pip install -U chromadb langchain-openai tiktoken
    ```

We must also add our imports and create our environment variables for API keys and LangChain configurations:

```
from langchain_openai import OpenAI, (ChatOpenAI,
    OpenAIEmbeddings)
. . .
# Environment variables for API keys and LangChain
configurations
os.environ['OPENAI_API_KEY'] = 'your-openai-api-key'
. . .
os.environ['LANGCHAIN_API_KEY'] = 'your-langchain-api-key'
```

2. Next, we must initialize `ChatOpenAI` and `OpenAIEmbeddings` so that we can generate responses and embeddings, respectively:

```
llm = ChatOpenAI(temperature=0.0, model="gpt-4-1106-preview")
embeddings = OpenAIEmbeddings()
```

3. Now, we can set up ChromaDB and our vector store. ChromaDB is an open-source database for managing embeddings, allowing us to store and search our embeddings. The vector store enables efficient retrieval of past conversation turns based on these embeddings. As we've created `PersistentClient`, notice that the Chroma database has been stored in an `emdeddings` folder under the current Jupyter Notebook folder:

```
chroma_client = chromadb.PersistentClient(path="./embeddings")
history_vectorstore = Chroma(
    persist_directory="embeddings/",
    collection_name="conversation_history",
    client=chroma_client,
    embedding_function=embeddings)
```

4. Now, we must create the memory and retriever objects. The memory object stores conversation turns, and the retriever helps us fetch relevant past turns by the vector store's default similarity search:

```
retriever = history_vectorstore.as_retriever(
    search_kwargs=dict(k=2))
memory = VectorStoreRetrieverMemory(
    retriever=retriever,
    memory_key="conversation_history",
    input_key="input")
```

5. Next, we must save some example conversations to memory so that we can check whether they're included in the context of the conversation later:

```
memory.save_context({"input": "Hi, my name is Adrian, how are
you?"}, {"output": "nice to meet you Adrian"})
memory.save_context({"input": "My favourite sport is running"},
{"output": "that's a good thing to be doing"})
```

6. Now, we must define a template for our conversation, which includes past conversations as context and the current input:

```
_DEFAULT_TEMPLATE = """
```

The following is a friendly conversation between a human and an AI sports assistant. The AI is helpful and provides specific details from its context to support the chat. If the AI doesn't know the answer to a question, it truthfully says it doesn't know:

```
Relevant pieces of previous conversation:
{conversation_history}
(Don't use these pieces of information if not relevant)
Current conversation:
Human: {input}
AI:"""
PROMPT = PromptTemplate(
    input_variables=["conversation_history", "input"],
    template=_DEFAULT_TEMPLATE
)
```

7. Finally, we must create `ConversationChain`, which combines all these components and sends a chat input:

```
conversation_with_history = ConversationChain(
    llm=llm,
    prompt=PROMPT,
    memory=memory,
    verbose=True)
response = conversation_with_history.invoke({
    "input":"what gear do i need to buy to get started in my
sport"})
print(response)
```

Here, `verbose` is set to `True` so that we can see the LLM's output. We'll see that the conversation history is included in the prompt. If we restart our Jupyter Notebook kernel and run the code again, the conversation should be persisted.

> **Tip**
>
> If you want to look at the documents and their embeddings stored in your Chroma vector store, you can output the whole collection by running `print(vector_store._collection.get(include=['embeddings']))`.
>
> You can also try out your own similarity search by running `similar_documents = vector_store.similarity_search(query="your question", k=1)`.

Summary

In this chapter, we have dived deeper into LangChain, looking at some more advanced topics. We started by looking at debugging techniques and introduced LangSmith, the go-to tool for advanced logging and monitoring for LangChain applications. We also looked at LangChain agents and understood how to use them to provide different functionalities to our LLM project, as well as how they tie in with OpenAI function calling. We focused on the out-of-the-box tools provided by LangChain and covered creating custom tools by looking at practical examples to help our ChatGPT application answer questions about real-time news and weather. Finally, we covered the concept of providing memory to our agents while looking at the different types of memory, the challenges, and the techniques involved in providing memory in LangChain.

In the next chapter, we'll look at RAG. You'll understand what it is, the concepts and processes involved, and how to implement retrieval in LangChain.

Further reading

The following links are a curated list of resources to help you understand this chapter:

- *LangChain*: `https://www.langchain.com`

  LangSmith: `https://smith.langchain.com`

  ChromaDB: `https://docs.trychroma.com`

- *Lost in the Middle: How Language Models Use Long Contexts*, by F. Liu: `https://arxiv.org/abs/2307.03172`

- *MemoryBank: Enhancing Large Language Models with Long-Term Memory*: `https://ar5iv.labs.arxiv.org/html/2305.10250`

- *Ebbinghaus's Forgetting Curve*: `https://www.mindtools.com/a9wjrjw/ebbinghaus-forgetting-curve`

  Virtual context management implementation: `https://memgpt.ai/`

- *LangChain agents*: `https://python.langchain.com/docs/modules/agents/`

  LangChain tools: `https://python.langchain.com/docs/integrations/tools/`

Part 3:
Building and Enhancing
ChatGPT-Powered Applications

In this part, we will explore **Retrieval-Augmented Generation (RAG)** systems, demonstrating how to use vector stores as a knowledge base. Then, we will bring together all the concepts covered so far to build and enhance ChatGPT-powered applications. You'll work on your own practical project to create a chatbot that can handle complex tasks and provide personalized interactions. Finally, we will look into the future of conversational AI and **large language models (LLMs)**, discussing the challenges of taking ChatGPT applications to production and examining alternative technologies and emerging trends in the field.

This part has the following chapters:

- *Chapter 7, Vector Stores as Knowledge Bases for Retrieval-augmented Generation*
- *Chapter 8, Creating Your Own LangChain Chatbot Example*
- *Chapter 9, The Future of Conversational AI with LLMs*

7

Vector Stores as Knowledge Bases for Retrieval-augmented Generation

Retrieval-augmented generation (**RAG**) is easily the most common use case for LLMs that has emerged since the explosion of ChatGPT. In this chapter, we're going to look at the key steps and concepts involved in creating a RAG system. Once you have an understanding of what's involved with each step, we'll look at how these processes and techniques can be carried out using LangChain. Going further, we'll work through our own RAG system with a real-world example.

This chapter aims to be an introduction to the core concepts of RAG so that you have a solid base for mastering it.

In this chapter, we'll cover the following topics:

- Why do we need RAG?
- Understanding the steps needed to create a RAG system
- Working through a RAG example with LangChain

Technical requirements

In this chapter, we will be using ChatGPT extensively, so you will need to sign up for a free account. If you haven't created an account, go to `https://openai.com/` and click **Get Started** at the top right of the page or go to `https://chat.openai.com`.

The examples require Python 3.9 and Jupyter Notebook to be installed, both of which can be found at `https://jupyter.org/try-jupyter/notebooks/?path=notebooks/Intro.ipynb`.

We'll use ChromaDB as our open source vector database, so it's advisable to take a look at it and familiarize yourself with Chroma (`https://www.trychroma.com`). However, there are a lot of other different options for vector store providers.

You can find this chapter's code in this book's GitHub repository: `https://github.com/PacktPublishing/ChatGPT-for-Conversational-AI-and-Chatbots/chapter7`.

Why do we need RAG?

LLMs are restricted by their knowledge of the world through their training data, so ChatGPT doesn't know about recent events or your own data, which severely restricts its ability to provide relevant answers. Things can also get worse with LLM performance because of hallucinations, where the LLM doesn't have any knowledge to support a question, so it makes things up.

So, when we talk about an LLM's knowledge, there are two types:

- Knowledge from information that the LLM used during training.
- Knowledge from information that was passed to the LLM via a prompt in the context of the conversation. We can call this context-specific knowledge.

So, the standout use case for an LLM application, and one that I'm asked about the most, is how we can allow an LLM to interpret and discuss data outside of their training dataset. This includes accessing real-time information or other external data sources, such as proprietary information contained within a company's document stores or databases. In effect, we need to improve our LLM's context-specific knowledge, and we do this with RAG.

Any requirement where the base language model may not have the breadth or depth of knowledge required means we need to source contextual information from somewhere so that we can pass it to our LLM. We also have to be mindful of the context window limitations of ChatGPT and the costs involved with larger context windows. If you have large amounts of data, then sending big chunks to ChatGPT is going to increase your token usage and costs. So, the smaller and more relevant our context-specific knowledge is, the better.

Although we will be focusing on the context-specific knowledge of other specific subjects for our ChatGPT application in this chapter, remember that context-specific knowledge can mean anything relevant to a conversation. You were introduced to the mechanics of a RAG system when we looked at persisting memory in the previous chapter. Let's look at the different elements of RAG in more detail.

Understanding the steps needed to create a RAG system

To implement RAG, a sequence of steps needs to be carried out. First, we need to define the sources from which data will be retrieved. This could range from online databases, specific websites, or even customized data repositories. Then, we need to optimize and store this information for retrieval. Once we've done this, we can use a retrieval system to fetch the relevant information based on the query context. This information is then passed on to the language model during the prompting process. The following figure outlines the different steps:

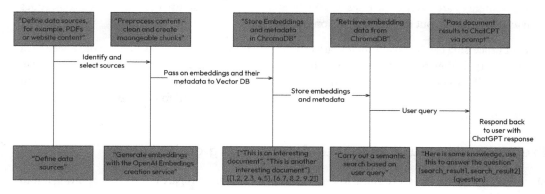

Figure 7.1 – Steps for creating a RAG system

I appreciate that a lot is going on here, so let's look at each step in more detail.

Defining your RAG data sources

For our first step, we must identify and select the sources from which our knowledge data will be retrieved. This could include online databases, specific websites, or customized data repositories tailored to the system's needs. Our sources could be in different types of file formats, such as raw HTML, PDF, PowerPoint, Word documents, or entire websites – it all depends on your use case.

To get the best results, you need to pre-process your data to ensure the best quality before starting the chunking process and other downstream tasks.

There is a lot you can do to improve your content before creating embeddings. This process can incur some costs, but it's worth the investment. Once you've done it, you can use the processed content in multiple downstream tasks.

Ideally, you don't want to care about where your content comes from concerning file type; you want to be able to process them regardless of source format so that you can treat them in the same way. To do this, you need to break them down into common elements such as titles and narrative text in a readable format. The most common format is JSON as it's a common and well-understood structure. This way, you can serialize content and extract and store metadata such as page number, document name, and keywords.

Every document type is different and employs different elements that we can use to differentiate between different types of information. For example, HTML has tags or we can use NLP to understand text; shorter sentences are likely to be titles. For PDFs and images, we can use more advanced techniques such as document layout detection or vision transformers.

Content processing can involve complex tasks and it's worth considering using a library or service such as Unstructured. They provide some really powerful core open source tools and great documentation, as well as APIs to carry out more complex tasks, such as content extraction. They also have an enterprise service in the pipeline. Take a look at their documents and follow their quickstart guide for more information.

Preprocessing our content and generating embeddings

Once we have the dataset we're going to use, we need to clean and prepare our data by splitting it into manageable chunks and creating embeddings from these chunks.

Chunking for effective LLM interactions

Your dataset will likely contain many documents and characters, so the first step of indexing your data is to get it into a reasonable size via text splitting. We do this for several reasons:

- To enhance result relevance by improving embedding accuracy for later processing.
- To support smaller context windows, which embedding models typically support.
- So that we can provide more accurate and specific information by narrowing down larger text sources into smaller, more focused segments.
- To reduce the volumes of contextual knowledge we're going to pass into the LLM. This keeps relevance up, which affects the quality of the response and keeps costs down. It also allows us to use larger documents.

When you chunk your text data, big parts of text are split into smaller chunks, such as sections, sentences, phrases, delimiters, and even words. With an effective chunking strategy, we aim to break down large content into meaningful pieces that we can use to accurately answer a user's query. The following figure highlights the different elements of a chunking strategy:

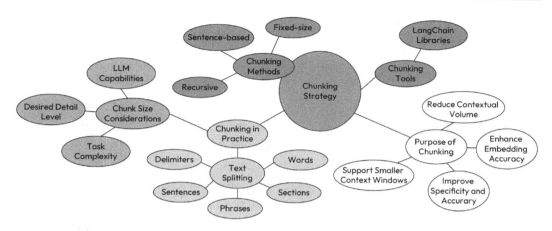

Figure 7.2 – Chunking strategy

However, chunking isn't a one-size-fits-all solution. Designing an effective strategy requires careful consideration of the content you're chunking and the use case for your application. Let's take a closer look at these elements:

- **Chunk size**: Think of this as the bite-size for your LLM. Smaller chunks (sentences, phrases) offer granular control but can lead to information loss and context fragmentation. Conversely, larger chunks (paragraphs, sections) maintain context but might overwhelm the LLM's capacity. Finding the sweet spot depends on the following factors:

  - **LLM capabilities**: Different embedding models have different handling capacities. Refer to the documentation for recommended context sizes. Don't forget that when the chunks are retrieved, we still need to add these to our context, so we also need to consider LLM context limits. Note that if you're using GPT-4 with the 32k content window, this is less of a problem.

  - **Task complexity**: What sort of content are you going to be using? Longer documents, such as articles, or smaller ones, such as chat transcripts? Simpler tasks, such as sentiment analysis, might handle larger chunks, while nuanced tasks, such as summarization, benefit from smaller bites.

  - **Desired level of detail**: Do you need a broad overview or a deep dive? What types of questions do you want to support? Tailor chunk size accordingly.

- **Chunking methods**: How you slice the text matters. Popular methods have different strengths and weaknesses, so I encourage you to try different approaches to see what sort of results you can achieve. All of these methods are supported by different libraries in LangChain. The following are some popular methods:

  - **Fixed-size chunking**: Simple and efficient but risks breaking sentences and losing context. We use fixed-size chunking the most in the examples in this book.

- **Sentence-based chunking**: Content-aware chunking allows you to split text into separate sentences.

- **Recursive chunking**: This chunking process iteratively splits text into smaller parts using varying criteria until chunks of a similar, though not identical, size are achieved, allowing for structural flexibility.

Remember, the optimal chunking strategy is unique to your LLM, task, and desired outcome. Experiment with different approaches, monitor performance, and refine your strategy to unlock the full potential of your LLM interactions.

The key takeaway is that LangChain provides some powerful out-of-the-box chunking tools. Now that we have our chunked data, we are ready to create our embeddings.

Understanding embeddings – transforming text into meaningful vectors

The terms **vector** and **embedding** are closely related in the context of machine learning and natural language processing, but they refer to different concepts. First, let's look at what a vector is:

- **What is a vector?**: A vector is a mathematical representation of data in a space defined by dimensions. It is essentially an array of floating-point numbers. Similar to packing items into a suitcase, we store various attributes of data, such as height, weight, color, or word frequency, in this list. This allows us to feed complex information into machine learning algorithms that only understand numbers. This enables efficient computation, allowing algorithms to handle large amounts of real-world data effectively.

- **Different types of vectors**: There are different types, such as feature vectors and embeddings (the vector type we are interested in for RAG).

- **They represent different things**: Vectors can represent various data types, from simple numerical features (such as image pixel values) to complex concepts such as word meanings or document topics.

- **Encoding with meaning**: Vector values aren't random; they hold meaning! For example, in word embeddings, the distance between words in the vector space reflects their semantic similarity.

- **From input to output**: Vectors can represent both input data fed into the algorithm and its output predictions. For instance, a vector (more on this later) might input some text as a vector and output a vector representing the closest answer to the question.

Word embeddings explained – from text to meaningful numbers

We can create various word or sentence embeddings. However, in the context of RAG, we are interested in chunked sentences, and we want to understand the relationship between them. To do this, we can use sentence embeddings.

Sentence embeddings transform entire sentences into dense vectors of fixed size, capturing the overall semantic meaning. Unlike word embeddings, which represent individual words, sentence embeddings consider the context and relationships between words in a sentence. We can measure the distance between vectors to show how related they are.

Let's imagine we're creating embeddings for the sentences "*I like eating cake*" and "*I'm interested in baking*" with a five-dimensional vector for each sentence. Note that real-world embeddings are much larger than this:

- Sentence: "*I like eating cake*:"

  - **Embedding**: `[0.45, -0.2, 0.33, 0.1, -0.4]`

- Sentence: "*I'm interested in baking*:"

  - **Embedding**: `[0.4, -0.15, 0.35, 0.05, -0.3]`

In this hypothetical example, each vector represents the entire sentence's meaning in a five-dimensional space. The dimensions are abstract and not directly interpretable but are structured in such a way that sentences with similar meanings are closer in this multidimensional space.

Imagine each sentence as a point in a giant room with five invisible walls (the dimensions). Sentences with similar meanings will be closer together in this room, even though you can't directly see those meanings.

Think of "*I like eating cake*" and "*I'm interested in baking*" as two people in the room. Sure, they're doing different things (eating versus baking), but they both share a love of food, kind of like being in the same "*food zone*" within the room. This is reflected in how close their points are, even though the room itself is invisible.

In real-world applications, this "*room*" would have many more walls (dimensions) to capture even finer details of meaning. The number of dimensions can range from 50 to 1,024, with 300 being a common choice.

The easiest way to create embeddings is to use a pre-trained model. We'll look at some of these models and how we can create embeddings in the next section.

OpenAI's text embedding models

OpenAI's text embedding models have consistently been among the most popular, with **text-embedding-ada-002** being the most widely used. However, these models have been superseded by even better embedding models. **text-embedding-3-small** and **text-embedding-3-large** are the newest and most performant embedding models.

OpenAI's latest advancement, Text Embeddings v3, represents a significant leap in artificial intelligence capabilities. This model, with its variations of "small" and "large," showcases notable improvements in performance, particularly in multilingual embeddings and English language processing. It introduces a more efficient approach to handling data dimensionality without compromising accuracy thanks to the innovative Matryoshka Representation Learning. This optimizes the embedding process by structuring data in layers, like Russian nesting dolls, which enhances the efficiency and accuracy of language models.

There are many other embedding models outside of OpenAI, such as Google's BERT and T5, Facebook's RoBERTa, and the multilingual XLM-R. The process of using different models is often streamlined by LangChain. For this chapter, we'll be concentrating on using the OpenAI embedding models. In the next section, we'll see how this looks.

Creating embeddings with OpenAI models

There are several different ways to use the OpenAI embedding models. One option is to call the embeddings API endpoint directly:

```
curl https://api.openai.com/v1/embeddings \
  -H "C OpenAI content-Type: application/json" \
  -H "Authorization: Bearer $OPENAI_API_KEY" \
  -d '{
    "input": ["I like eating cake"," I'm interested in baking"],
    "model": "text-embedding-3-small"
  }'
```

Alternatively, LangChain provides an easy OpenAIEmbeddings wrapper class to create embeddings with the OpenAI API. Here's a simple example:

```
import ...
os.environ['OPENAI_API_KEY'] = 'KEY'
embeddingsService = OpenAIEmbeddings()
embeddings = embeddingsService.embed_documents(["I like eating cake",
    " I'm interested in baking"])
```

Often, there's no need to directly use an embedding creation service. Many vector stores offer a streamlined approach to both embed and store documents simultaneously. We will illustrate this by examining the process within ChromaDB. Now that we've covered vectors and embeddings, we can move on to the next step in our RAG process: storing and searching our embeddings with a vector store.

Storing and searching our embeddings with a vector store

Now that we have our high-dimensional data vector embeddings, we need to be able to store them and, more importantly, search them effectively. Our vector database is going to play an important role in storing and searching our data. The following diagram illustrates this process:

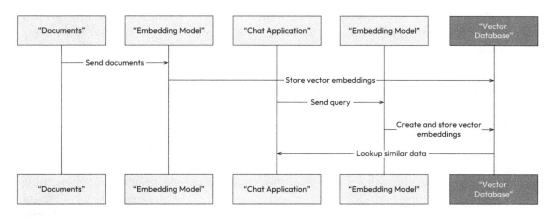

Figure 7.3 – Indexing our data

Our vector database is a specific type of database that's been optimized to work with multi-dimensional vectors. Let's look at the advantages of using this type of storage.

Main advantages of vector databases

Vector databases are purpose-built to manage vector embeddings, so they provide several advantages:

- **Straightforward data management**: Vector databases provide all the **Create, Read, Update, Delete (CRUD)** actions we expect with a database. Vector databases also provide simple ways to carry out metadata storage and filtering so that we can query using additional metadata filters for finer-grained queries.

- **Similarity search**: The main advantage of using a vector database is that it allows us to search our embedding based on the similarity of the meanings represented in our vectors. So, while a traditional database is queried based on exact matches or criteria, we can use a vector database to find the most similar or relevant data based on semantic or contextual meaning.

- **Scalability**: With most use cases, we'll likely be working with large volumes of vectors. As the data volume grows, vector databases can scale horizontally by distributing data across multiple machines, ensuring efficient performance even with massive datasets.

How does a vector database work?

While traditional databases store data in rows and columns, accommodating scalar data types such as strings and numbers, vector databases are designed around managing vectors. This fundamental difference extends to their optimization and querying mechanisms.

In a conventional database setup, queries typically seek exact matches among rows based on specified criteria. Vector databases, however, employ similarity metrics to identify vectors closely resembling the query, diverging from the exact-match approach.

At the core of vector databases is the **approximate nearest neighbor** (**ANN**) search, which leverages a suite of algorithms to refine and accelerate the search process. These algorithms, which include methods such as hashing, quantization, and graph-based searches, are orchestrated into a pipeline to return fast, accurate results for the query.

Given that vector databases yield approximate results, a key consideration is the balance between accuracy and query speed. Higher accuracy levels often result in slower searches. Nevertheless, a well-optimized vector database can achieve remarkably fast searches without significantly compromising accuracy, striking an optimal balance for many applications. Vector databases employ several different techniques or similarity measures for comparing vectors within a database so that they can identify the most relevant outcomes.

Similarity measures are mathematical techniques that are used to quantify the resemblance between two vectors within a vector space. This enables the database to assess the similarity of stored vectors against a query vector. There are several similarity measures, with the following being some standout ones:

- **Cosine similarity**: This measure calculates the cosine of the angle between two vectors, with values ranging from -1 to 1. Here, a value of 1 indicates identical vectors, 0 denotes orthogonal (independent) vectors, and -1 suggests completely opposing vectors.

- **Euclidean distance**: This measures the direct line distance between two vectors, extending from 0 to infinity. A value of 0 signifies identical vectors, with increasing values indicating growing dissimilarity.

- **Dot product**: This represents the multiplication of the magnitudes of two vectors and the cosine of the angle between them. Positive values indicate vectors pointing in the same direction, 0 denotes orthogonal vectors, and negative values suggest opposite directions.

Now that you understand the basics of how vector databases function, the next step is to choose the right one for your project.

Deciding which vector database to use

Many vector databases and vendors provide open source, hosted, scalable solutions catering to every price point and requirement

Like selecting any persistence layer, choosing the right vector database for RAG systems and LangChain applications requires carefully considering several key factors tailored to these advanced AI frameworks.

For RAG systems, the vector database must excel in fast and accurate retrieval capabilities, and it must be easy to work with. Vector databases often provide programming-language-specific SDKs that wrap the API, which will make it easier for you to interact with the database in your applications.

In the context of LangChain, most of the main vector providers have already created integrations in the LangChain library and are part of the `langchain-community` packages. Just have a look at the integrations pages to find out which vendors have created out-of-the-box integrations and extensive documentation.

Budget considerations also play a critical role; open source vector databases might offer cost savings and customizability for teams willing to invest in setup and maintenance. However, for projects requiring seamless scalability and minimal operational overhead, a managed cloud-based vector database could be more appropriate, despite potentially higher costs. The scope and scale of a project are important here; it may be that you have the luxury to choose your provider or it may be that you're required to look at technologies on offer from your cloud provider of choice.

Standout vector database providers

When it comes to vector databases, there are both open source and managed options available. Let's look at a few examples:

- ChromaDB is one of the solid open source vector databases. We've already used this in this book, and we'll use it as our go-to vector database for the rest of this book. Chroma is an open source embedding database that allows you to easily manage text documents, convert text into embeddings, and do similarity searches. Yes, there is a ChromaDB LangChain wrapper and retriever classes for querying. Other standout open source vector stores worth mentioning are Qdrant and Weaviate.

- **Pinecone** is a managed vector database platform that's been purpose-built to tackle the unique challenges associated with vectors and provides everything you need to go into production.

- **Facebook AI Similarity Search** (**FAISS**), provided by Meta, is an open source library for swiftly searching for similarities and clustering dense vectors. It's not exactly a vector database; it's a vector index, also known as a vector library. It houses algorithms that are capable of searching within vector sets of varying sizes and complexity. FAISS is well supported in LangChain.

These are just a few of the vector database options available. I highly recommend conducting thorough research to determine the most suitable choice for your project. Additionally, consider exploring the vector providers featured within the LangChain integrations for a more comprehensive understanding of what's available.

At this point, we understand the sequence of steps in our RAG. We've learned how to define our data sources, preprocess our data into chunks, and create our embeddings with specialist models, as well as how we leverage vector databases to store and search. In the next section, we'll look at an end-to-end example that brings all these techniques together using the LangChain framework.

Working through a RAG example with LangChain

LangChain provides functionality to carry out all of the steps we've outlined. So, let's look at a RAG example while looking at how we achieve the steps with LangChain in more detail.

For our use case, we're going to look at using unstructured website data as the basis for our RAG system. This is a common example of a RAG application as most organizations have websites and unstructured data that they want to use. Imagine that your organization has asked you to create an LLM-powered chatbot that can answer questions about the content on your organization's website.

In our scenario, we'll explore leveraging unstructured website data as the foundation for our RAG system. Utilizing unstructured data from websites is a prevalent approach for RAG applications given that most organizations possess websites filled with data they wish to use. Imagine being tasked by your organization to develop a chatbot capable of answering queries related to the content on your company's website.

For our example, we're going to create a chatbot that can provide up-to-date information about New York's economy. For our data source, we'll use Wikipedia. You're welcome to follow your own subject matter for this and even use your own website pages.

The first step is to load our dataset with a document loader. So, let's dive in.

Integrating data – choosing your document loader

LangChain document loaders are designed to easily fetch documents from different sources, serving as the initial step in the data retrieval process. LangChain provides access to over 100 different document loaders that cater to a diverse range of data sources, such as private cloud storage, public websites, or even specific document formats such as **HTML**, **CSV**, **PDF**, **JSON**, and code files.

The document loader you need will depend on your use case and requirements. With the options available, you can integrate data from practically any source into your ChatGPT application, whether that be unstructured public or private data from websites or cloud storage such as Google Drive. You can also load more structured information from many different database types or data storage providers.

In addition to their versatility, document loaders in LangChain are designed to be easy to use. They offer straightforward interfaces that allow developers to quickly set up and start fetching documents with minimal configuration. This simplicity accelerates the development process, particularly when you're creating proofs of concept to try out different data sources. Hopefully, this will give you more time to focus on more complex aspects of your application.

Let's look at a simple example of a document loader. Remember that we want to answer questions about New York, so our dataset needs to match the requirements. Wikipedia is a great source for up-to-date information, so this will suffice for this example. Let's look at using `WebBaseLoader`, an **HTML** document loader, to pull some information from Wikipedia as the basis for our dataset:

```
from langchain_community.document_loaders import WebBaseLoader
loader = WebBaseLoader(
    ["https://en.wikipedia.org/wiki/New_York_City#Economy"])
docs = loader.load()
print(len(docs))
print(docs)
```

Printing out the `docs` variable should result in something similar to the following, where `page_content` is the text data we're going to use:

```
[Document(page_content='\n\n\n\nNew York City - Wikipedia
...
 metadata={'source': 'https://en.wikipedia.org/wiki/New_York_
City#Economy', 'title': 'New York City - Wikipedia', 'language':
'en'})]
```

Considering the unstructured HTML data we have, it may also be worth considering using `documentTransformer` to clean up our HTML into plain text, such as `AsyncHtmlLoader`.

Many other LangChain document loader options are purpose-built for specific use cases, so it's always worth checking what's available in Python or TypeScript.

For our example, we'll use `WikipediaLoader`, which we used earlier in this chapter. Note that I'm keeping `load_max_docs` low for the interests of this example:

```
docs = WikipediaLoader(query="New York economy",
    load_max_docs=5).load()
```

The benefit of using this is that we now have data from multiple Wikipedia pages. The results from `WikipediaLoader` should look pretty good, with a list of documents similar in structure to the `WebBaseLoader` results.

Now that we have our dataset, the next step is to index it.

Creating manageable chunks with text splitting

As we discussed in the section on Chunking for Effective LLM Interactions, the process of text splitting can be quite complicated, requiring careful decisions about chunk size and the appropriate method to use. Fortunately, LangChain provides easy-to-use interfaces and integrations for several modules that can be used to "chunk" or split our document dataset so that it's ready for further processing.

LangChain's text-splitting algorithms are designed to be versatile and efficient, ensuring that the integrity and context of the original document are maintained while optimizing it for retrieval. This involves considering factors such as the nature and format of the text, the logical divisions within the document, and the specific requirements of the retrieval task at hand.

Implementing text splitting in LangChain can be done by using one of the out-of-the-box splitters, which provide different ways to split depending on the dataset itself.

> **Tip**
> One fantastic tool for understanding different chunking strategies and evaluating the results of chunking your text is `https://chunkviz.up.railway.app`. It's a brilliant way to see how your text is split based on the LangChain utility and different parameters such as chunk size and overlap.

For this use case, we're going to use `RecursiveCharacterTextSplitter` to chunk our text. The `RecursiveCharacterTextSplitter` utility is designed to efficiently segment text into manageable pieces. It looks for predefined separator characters such as newlines, spaces, and empty strings. The result should be chunked text broken down into paragraphs, sentences, and words, where the text segments preserve the meaning of related text segments. The utility dynamically divides text into chunks, targeting a specific size in terms of character count. Once a chunk reaches this predetermined size, it's earmarked as an output segment. To ensure continuity and context are maintained across these segments, adjacent chunks feature a configurable degree of overlap.

Its parameters are fully customizable, offering flexibility to accommodate various text types and application requirements.

It's a good solution for breaking down large volumes of text into smaller, semantically meaningful units.

Here is our example code, which uses `RecursiveCharacterTextSplitter`. This is recommended for generic text:

```
from langchain.text_splitter import RecursiveCharacterTextSplitter
text_splitter = RecursiveCharacterTextSplitter(
    chunk_size=500, chunk_overlap=100)
split_documents = text_splitter.split_documents(docs)
print(len(split_documents))
```

We can create our `RecursiveCharacterTextSplitter` class with attributes that we wish to control, including `chunk_size`, which ensures manageable sizes for processing, and `chunk_overlap`, which allows for character overlap between chunks, maintaining context continuity. If you change the chunk size, you'll be able to create a document list with more entries. You can look at one of the split documents by running the following command:

```
print(documents[5])
```

With `RecursiveCharacterTextSplitter`, we've broken down our large text content into smaller, meaningful units. This prepares our data for further processing and analysis. Now that we've explored text splitting, let's delve into how we can use our chunks to create embeddings:

Creating and storing text embeddings

Now that we have chunks of manageable data, it's time to create embeddings from these chunks and store them in our vector database. For this example, we're going to use ChromaDB as our vector store.

Here, we're going to create our embeddings from the `langchain_openai` and `langchain.vectorstores` libraries, which allow us to call the OpenAI embeddings service and store them in ChromaDB. Follow these steps:

1. First, we must initialize an instance of the `OpenAIEmbeddings` class with our specific model, `text-embedding-3-small`:

    ```
    from langchain_openai import OpenAIEmbeddings
    from langchain.vectorstores.chroma import Chroma
    embeddings_model = OpenAIEmbeddings(
        model=" text-embedding-3-small")
    ```

2. The `docs_vectorstore = Chroma(...)` line creates an instance of the Chroma class, which is used to store document vectors. The following parameters are provided to Chroma:

    *   `collection_name="chapter7db_documents_store"`: This specifies the name of the collection within Chroma where the document vectors will be stored. Think of it as naming a table in a database where your data will be kept.

    *   `embedding_function=embeddings_model`: Here, the previously created `embeddings_model` is passed as the embedding function. This tells Chroma to use the `OpenAIEmbeddings` instance to generate embeddings for documents before storing them. Essentially, whenever a document is added to Chroma, it will be passed through this embedding model to convert the text into a vector.

- `persist_directory="chapter7db"`: This parameter specifies the directory where Chroma will persistently store its data. This allows the vector store to maintain its state between program runs. In our case, it will be a folder where we're running our Jupyter Notebook:

```
documents_vectorstore = Chroma(
    collection_name="docs_store",
    embedding_function=embeddings_model,
    persist_directory="chapter7db",
)
```

3. We can then pass in our previously split documents by calling `add_documents()`. After, we can persist everything by calling `persist()` on our Chroma instance:

```
chroma_document_store.add_documents(split_documents)
chroma_document_store.persist()
```

4. Now, we can conduct a search to see if we get any results back. We'll do this by carrying out a similarity search on the Chroma instance by calling `similarity_search` with our text query. Try a question related to New York's economy:

```
similar_documents = chroma_document_store.similarity_search(
    query="what was the gross state product in 2022")
print(similar_documents)
```

You should be able to print out search results for your query. Now, let's look at bringing our RAG system together with LangChain.

Bringing everything together with LangChain

Now that we have our Chroma document store returning results, let's learn how to bring all our RAG elements together using LangChain so that we can create a system that's capable of answering questions based on information in our documents:

1. We'll begin by importing the necessary classes from the `langchain_openai` and `langchain_core` libraries, as well as the variables we need to log in to LangSmith. Then, we need to create an instance of the `ChatOpenAI` class so that we can leverage GPT-4 to generate our answers:

```
from langchain_openai import ChatOpenAI
from langchain_core.prompts import ChatPromptTemplate
from langchain_core.runnables import (
    RunnableParallel, RunnablePassthrough)
os.environ['LANGCHAIN_API_KEY'] = ''
os.environ['LANGCHAIN_PROJECT'] = "book-chapter-7"
os.environ['LANGCHAIN_TRACING_V2'] = "true"
os.environ['LANGCHAIN_ENDPOINT'] = "https://api.smith.langchain.
com"
```

```
from langchain_core.output_parsers import StrOutputParser
llm = ChatOpenAI(model="gpt-4", temperature=0.0)
```

Setting the temperature to 0.0 makes the model's responses deterministic, which is often desirable for consistent QA performance.

2. Now that our Chroma document store has been set up, we can initialize a retriever to fetch relevant documents based on our query:

```
retriever = chroma_document_store.as_retriever(
    search_kwargs={"k": 5})
```

The retriever is configured to return the top 5 documents that match a query.

3. Next, we need to create a prompt template that tells the language model how to utilize the retrieved documents to answer questions:

```
template = """
You are an assistant specializing in question-answering tasks
based on provided document excerpts.
Your task is to analyze the given excerpts and formulate a final
answer, citing the excerpts as "SOURCES" for reference. If the
answer is not available in the excerpts, clearly state "I don't
know."
Ensure every response includes a "SOURCES" section, even if no
direct answer can be provided.
QUESTION: {question}
=============================
EXCERPTS:
{chroma_documents}
=============================
FINAL ANSWER(with SOURCES if available):
"""
prompt = ChatPromptTemplate.from_template(template)
```

This template helps the model provide sourced answers from our data or instructs us to indicate when information is unavailable:

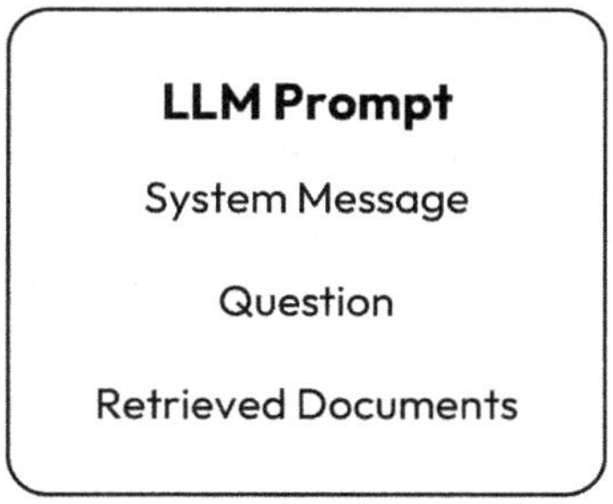

Figure 7.4 – Prompt passed to the LLM

4. Next, implement a function to format the retrieved documents so that they can be included in the prompt:

```
def format_chroma_docs(chroma_docs) -> str:
    return "\n".join(
        f"Content: {doc.page_content}\nSource: {doc.
metadata['source']}" for doc in chroma_docs
    )
```

We'll use this function to organize document content and sources into a readable format when they're included in our prompt.

5. Now, we must use LCEL to create our question-answering pipeline. This setup allows us to directly apply formatting to the retrieved documents and streamlines the flow into the prompt generation and answer production stages:

```
rag_chain_from_docs = (
    {"chroma_documents": retriever | format_docs,
        "question": RunnablePassthrough()}
    | prompt
    | llm
    | StrOutputParser()
)
```

Here's what each part of this pipeline does:

I.   Processing with `RunnablePassthrough` and document retrieval: `{ "chroma_documents": retriever | format_docs, "question": RunnablePassthrough() }`: This dictionary sets up two parallel streams. For `chroma_documents`, documents are retrieved via `retriever` and then immediately formatted with `format_docs`. For `question`, `RunnablePassthrough()` is used to simply pass the question through without any modification. This approach efficiently prepares both the formatted documents and the question for the next step.

II. Prompt application: `| prompt`: The processed information (both question and formatted documents) is then piped into the prompt. This step utilizes `ChatPromptTemplate` to combine the question with the formatted documents according to the predefined template, setting the stage for the LLM to generate an answer.

III. Language model generation: `llm`: Here, the combined input from the previous step is fed into the language model (GPT-4) initialized earlier. The model generates a response based on the prompt, effectively answering the question with the help of information extracted from the retrieved documents.

IV. Output parsing: `| StrOutputParser()`: Finally, the output from the language model is parsed into a string format by `StrOutputParser()`. This step ensures that the final answer is easily readable and formatted correctly.

6. Finally, we can invoke the pipeline with a question and print the answer:

```
question = "what stock exchanges are located in the big apple"
rag_chain_from_docs.invoke(question)
```

This search should return something similar to the following, including an answer and sources:

```
The stock exchanges located in New York City, often referred
to as the "Big Apple", are the New York Stock Exchange and
Nasdaq. These are the world\'s two largest stock exchanges by
market capitalization of their listed companies. \n\nSOURCES:
\n- https://en.wikipedia.org/wiki/New_York_City\n- https://
en.wikipedia.org/wiki/Economy_of_New_York_City'
```

By following these steps, you've constructed a RAG system that's capable of answering questions about New York's economy with up-to-date information from Wikipedia. Try some other questions and see how the FAQ performs.

Remember that it's useful to look at the debugging output in LangSmith so that you can see what's going on in each step of LangChain. In the following screenshot, we've drilled down into the LangSmith prompt template output so that we can see the documents we've retrieved and passed into the LLM prompt:

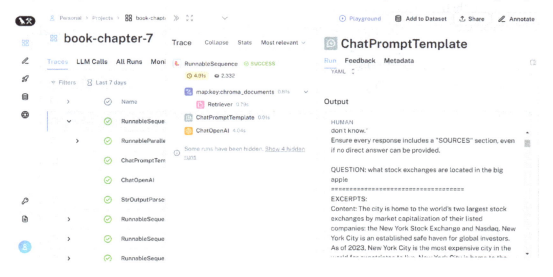

Figure 7.5 – Using LangSmith to see what's going on under the hood

First, we loaded our documents from Wikipedia based on our subject criteria, chunked this data, and created embeddings that we've stored in our vector store. Then, we defined a retriever to query our store, passed the document search results to an LLM prompt, and handed over to our LLM to get an accurate response to our question while listing sources and safeguarding against hallucinations. This was brought together with a LangChain defined in LCEL.

Summary

In this chapter, we looked at what RAG systems are and why we need them. Then, we covered the steps we must follow to create a RAG system in detail. At this point, you should know how to select and pre-process data with chunking, as well as what vectors and embeddings are and how to create them with specialized models so that you can store them with a specific type of database for storing vectors. Finally, we created a RAG system, working through each step with a real-world example before bringing it all together with LangChain.

In the next chapter, we'll look at a more complex ChatGPT-powered, real-world LangChain application in more detail. We'll be utilizing some of the core concepts from previous chapters as well as what we've learned about how to implement RAG in this chapter. We'll aim to create a more complex chatbot that supports memory, RAG, and more specific capabilities powered by LangChain tools.

Further reading

The following is a curated list of resources to help you understand the material provided in this chapter:

- LangSmith: `https://smith.langchain.com/`
- Embedding models: `https://python.langchain.com/docs/integrations/text_embedding` and `https://platform.openai.com/docs/guides/embeddings/embedding-models`
- ChromaDB: `https://docs.trychroma.com/`
- Unstructured: `https://docs.unstructured.io/welcome`

8

Creating Your Own LangChain Chatbot Example

In this chapter, we'll combine key concepts from earlier chapters into a practical project. So far, we've covered a lot, from conversational AI fundamentals and prompt engineering to utilizing LangChain to create a RAG system. Now, we're ready to apply this knowledge by building a ChatGPT-powered chatbot.

This project will demonstrate how to apply ChatGPT technology in a practical, real-world conversational AI scenario. We'll aim to create an automated assistant that can not only answer questions about our own data but also handle more complex tasks and provide personalization.

By the end of this chapter, you'll not only have a functional ChatGPT chatbot but also a solid understanding of how to craft more sophisticated conversational agents. You'll also gain insights to better assess the feasibility of integrating LLM-powered agents into your business operations and determine whether it's the right moment to transition from existing NLU systems.

In this chapter, we'll cover the following key areas:

- Scoping our ChatGPT project
- Getting our data ready for the Chatbot
- Creating our agent for complex interactions
- Bringing it all together – building your own LangChain chatbot with Streamlit

Let's start building our chatbot, but first, we need to consider our conversational agent's use case and scope.

Technical requirements

In this chapter, we will use ChatGPT extensively, so you will need to be signed up with a free account. If you haven't created an account, go to `https://openai.com/` and click **Get Started** at the top right of the page, or go to https://chat.openai.com.

The examples require Python 3.9 and a Jupyter notebook to be installed: `https://jupyter.org/try-jupyter/notebooks/?path=notebooks/Intro.ipynb`.

We will use Chroma DB as our open source vector database, so it's advisable to take a look and familiarize yourself with Chroma (`https://www.trychroma.com`) although there are a lot of other different options for Vector stores providers.

Scoping our ChatGPT project

Everyone likes a holiday, and helping travelers with smart conversational automations is an area I'm very passionate about, so for our use case, we are going to create our own ChatGPT-powered travel assistant.

At this point, we must be realistic with scope; an online travel agent could potentially do thousands of different tasks with varying levels of complexity, but let's choose a handful of specific use cases that are not only interesting to implement but also bring together the techniques we've covered so far in the book.

A holiday assistant use case

You work for an online travel agent, and you've been tasked with creating an LLM-powered holiday assistant designed to revolutionize the way travelers choose their accommodations. Online travel agents tend to be limited in their accommodation searches, so your challenge is to provide a more personalized granular search. Drawing on a dataset of hotels, complete with descriptions, reviews, and other possible sources of information such as weather and location, your goal is to construct a ChatGPT-powered chatbot capable of guiding users to their ideal hotel choice. Let's call our agent Ellie the explorer and dig deeper into her persona.

A persona outline for Ellie the explorer

First, let's consider Ellie's personality and the capabilities she's going to offer to her users.

Ellie's personality

Ellie is enthusiastic, friendly, and incredibly knowledgeable. She approaches every interaction with a positive attitude and a genuine desire to help. Her responses are thoughtful and personalized, reflecting her understanding of each user's unique preferences and needs. Ellie is not just a chatbot; she's a travel companion who's excited to help users plan their trips, offering advice as a friend would.

Ellie's interactions are characterized by her warmth and enthusiasm. She communicates in a conversational tone, making complex travel planning feel like a chat with a knowledgeable friend. Her responses are prompt, informative, and always considerate of the user's needs and preferences.

Ellie's capabilities

Let's outline what Ellie's capabilities are going to be when handling her users' queries:

- **Learning user preferences**: Ellie is attentive and observant, quickly picking up on subtle hints about a user's travel preferences, family dynamics, and favorite holiday activities. She remembers these details across interactions, continually refining her understanding of what each user enjoys.

- **Personalized recommendations**: Using her comprehensive knowledge of a hotel's dataset, Ellie goes beyond simple keyword searches. She understands free text questions and digs deep into her database to offer recommendations that perfectly match each user's wishes, whether they're looking for family-friendly resorts, romantic getaways, or adventure-packed hostels.

- **Destination information**: Ellie is a treasure trove of information on countless destinations. She can tell you the best times to visit a place, share insights on local attractions, offer safety advice, and even teach you a few cultural tips to help you blend in. Her advice is always up to date, ensuring travelers have the best experience.

- **Weather conditions**: Ellie keeps a close eye on the weather across the globe. She can provide current and forecasted conditions for any destination, advising on how the weather might affect travel plans. Whether it's suggesting the best day for a beach outing or warning about potential travel disruptions due to severe weather, Ellie's guidance is invaluable.

Now we've got an idea about Ellies capabilities, let's look at her conversational capabilities and the types of interactions we want to support.

Ellie's conversational scope

Let's outline the conversations Ellie will be able to carry out. She will hopefully be able to carry out the following types of interactions:

- **Personalized recommendations**: Using the hotel's dataset, based on user preferences and past interactions, the chatbot can offer personalized hotel recommendations, tailored to meet each user's specific needs. We want to offer more than just keyword searches for parameters such as facilities. We want to support free text questions about our hotels based on hotel descriptions and holiday maker reviews, allowing us to offer more granular suggestions to our users.

- **Destination information**: Inquiries about the destination, including best times to visit, local attractions, safety advice, and cultural tips.

- **Weather conditions**: Questions about current and forecasted weather conditions for specific destinations or how the weather might affect travel plans.

Hopefully, once our **proof of concept(POC)** is working, we'll be able to make an informed decision on whether we want to make a more fully featured agent with a larger hotel dataset.

Technical features

To create Ellie the explorer and meet the conversational scope we've outlined, we're going to implement the following features:

- A RAG system to store information about our hotel descriptions that we can use to provide our suggestions

- LangChain agents to provide the tools to interact with external data sources and get up-to-date information

- Performance monitoring using LangSmith so that we can see how our assistant performs

- A working web-based Chatbot interface where we can see our conversations and interact with the agent, while keeping track of the ongoing conversation

Getting our data ready for the chatbot

To get our data ready, we first need to source and then process it so that it's in the cleanest state, ready to be used.

Selecting our data sources

For this project, we're going to need to provide a number of different types of context-specific knowledge to our LLM agent so that we can service the expected question types:

- **Hotel information**: This would come from our online travel agent hotel dataset, so we'll need some hotel data to represent the hotels we want to recommend to our users. The data we're going to use to provide hotel recommendations is a small subset of the dataset at `https://www.kaggle.com/datasets/raj713335/tbo-hotels-dataset`. This dataset contains information on 1,000,000+ hotels from different countries and regions, such as their rates, reviews, amenities, location, and star ratings. The data was collected from various sources, such as hotel websites, online travel agencies, and review platforms. We're going to use a much smaller version of this data to keep our model interaction costs down.

- **Location weather**: Users may ask about the weather in a specific location, so we'll use Open-Meteo, the free weather API.

- **Detailed location information**: To provide up-to-date location information, we'll use Wikipedia so that we can search by location.

Preparing the hotel data

The hotel dataset we're going to use, `hotel-info-chapter8-001.csv`, is a **CSV** file made up of around 900 hotel records from countries around the world for a bit of variety. This data is populated with the following columns:

`countryCode`, `countryName`, `cityCode`, `cityName`, `HotelCode`, `HotelName`, `HotelRating`, `Address`, `Attractions`, `Description`, `HotelFacilities`, `Map`, `PhoneNumber`, `PinCode`, `HotelWebsiteUrl`, and `Reviews`

These are all fairly self-explanatory; the columns we're most interested in are the `Description` and `Reviews` columns, as we're going to use this data to form the basis for our RAG system to help Ellie provide hotel recommendations. So, to recap, what we need to do to make this data available for RAG is to *clean it up, chunk and create embeddings, and store the data as embeddings so that we can retrieve it later on?.

Cleaning and normalizing data

Before creating embeddings, we need to ensure that our data in the `Description` and `Reviews` columns is clean and normalized. This involves removing any HTML. If you look at the `descriptions` field, you'll notice that it has text with HTML tags, so the first thing we should do is clean this up. To do this, we can use the `BeautifulSoup` library from Python's `bs4` package. In the following code, we'll any HTML in `Descriptions` and `Attractions` with the `remove_html_tags` function:

```
def remove_html_tags(text):
    if pd.isna(text):
        return ""  # Return an empty string for NaN values
    elif isinstance(text, str):
        return BeautifulSoup(text, "html.parser").get_text()
    else:
        # Convert non-string, non-NaN values to string
        return str(text)
df['Description'] = df['Description'].apply(remove_html_tags)
df['Attractions'] = df['Attractions'].apply(remove_html_tags)
```

Other improvements to our cleaning process could involve standardizing abbreviations (e.g., changing `Rest.` to `Restaurant`) and correcting spelling errors. Normalizing data helps to reduce the variance in your embeddings, leading to more accurate recommendations.

Creating our chunks

Let's look at creating our chunks from the `Description` and `Reviews` columns. There are a number of different ways you could do this; we're using a LangChain document loader, but in a production environment where you could have a much larger dataset, you may find using the `pandas` library we used to clean up our data in the previous section is more efficient. In our approach, we read the data out of our CSV file, create a dictionary from each row so that we can access the column values, look at each document, and then create chunks from the `Description` and `Reviews` columns:

1. **Load the CSV file**: First, we need to load our hotel information from a CSV file. We'll use a custom `CSVLoader` class for this purpose. Specify the file path, the column containing hotel names, and any additional metadata columns. We're using the `HotelName` as the source column:

```
loader = CSVLoader(
    file_path="hotel-info-chapter8-001_cleaned.csv",
    source_column="HotelName",
    metadata_columns=["countyCode", "countyName",
        "cityCode", "cityName", "HotelCode", "HotelName",
        "HotelRating", "Address", "Attractions",
        "HotelFacilities", "Map", "PhoneNumber",
        "PinCode", "HotelWebsiteUrl"]
)
docs = loader.load()
```

2. **Create a text splitter**: To handle potentially long hotel descriptions and reviews, we'll split them into smaller chunks. We use `RecursiveCharacterTextSplitter` for this, setting a chunk size and overlap to ensure continuity in the text analysis. The descriptions are not actually that large, but for our example, the aim is to break them down into small enough chunks to provide some granular information about each hotel:

```
splitter = RecursiveCharacterTextSplitter(
    chunk_size=250, chunk_overlap=20)
```

3. **Process and split the text**: Iterate through each document, extract the content of the specified columns, and split the content of each column into manageable chunks. We then loop over each chunk and create a document for each chunk, consisting of a key/value pair of `metadata` and `page_content`, ready for downstream processing:

```
chunked_docs = []
for doc in docs:
    content_dict = {}
    pairs = doc.page_content.split('\n')
    for pair in pairs:
        if ': ' in pair:
            key, value = pair.split(': ', 1)
            content_dict[key.strip()] = value.strip()
    for column in content_columns:
        if column in content_dict and content_dict[column]:
            chunks = splitter.split_text(content_dict[column])
            chunked_docs.extend(
```

```
[
    Document(page_content=doc_part,
        metadata=doc.metadata)
    for doc_part in chunked_doc_parts
]
)
```

We want to loop over each CSV record and select `Description` and `Review`, which are then chunked individually, so each chunk with the original CSV metadata is then added to our `chunked_docs` list ,ready to create embeddings from.

Creating our embeddings

Once we've got our chunked text, we want to create our embeddings. For our vector database, we're going to use Chroma DB and the OpenAI model to create our embeddings. We'll store the database locally and use the latest OpenAI embeddings model, `text-embedding-3-small`, to create our embeddings. To keep my OpenAI embedding costs down, I'm only going to create embeddings from the first 100 chunks; it's up to you how many you want to create for your project:

```
embeddings_model = OpenAIEmbeddings(
    model="text-embedding-3-small")
chroma_document_store = Chroma(
    collection_name="chapter8db_hotel_store",
    embedding_function=embeddings_model,
    persist_directory="chapter8db",
)
chroma_document_store.add_documents(chunked_docs[:100])
```

We can test whether we've got records in the store by looking at the number of records with `count()` or doing a similarity search:

```
print("There are",
    chroma_document_store._collection.count(),
    "in the collection")
similar_documents = chroma_document_store.similarity_search(
    query="I'm interested in going trail running",k=1)
similar_documents
```

Now we've got our hotel embeddings stored with their metadata, we can start to look at creating the tools that our agent will utilize to interact with this data and other sources.

Creating our agent for complex interactions

Our agent is going to be powered by tools that provide the functionality to meet our requirements. The main concept is that the agent will be able to take our question and use an LLM call to reason and split this task down into smaller tasks, while deciding which tools are appropriate to use to satisfy our request and deliver an answer to our question. The reasoning processes the agent uses are illustrated in the following diagram:

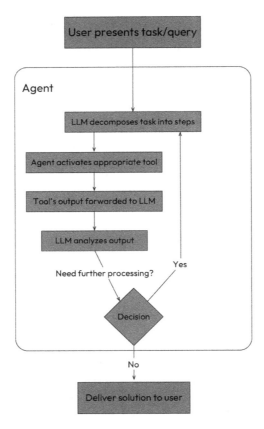

Figure 8.1 – The agent's reasoning process

Begin by creating a dedicated folder that will contain all the code for our agent. This will help organize our project files and make the code base easier to navigate. I'm going to name this folder `streamlit`, as it's where we'll create our Streamlit app later on.

Copy over the Chroma database we've created to store the vectors in this folder: `path to your project\streamlit\chapter8db\`.

Next, create a folder named `tools` so that we can start creating the tools for our agent.

Creating the agent tools

Before we create our LangChain agent, we'll need to create the LangChain tools to meet our requirements. We're going to try and implement all our functionality using agent tools, including the RAG functionality:

- Hotel information
- Weather information
- Location information

Each tool will have the capabilities to carry out its own task. Let's work through each one; for now, create a folder named `tools`, where we'll save the tools to use in our final project.

Personalized hotel recommendations

The first tool is arguably the most important for our application, as this is the tool that will interact with our vector store that makes up the retrieval part of our RAG system. We're going to create a custom tool to accomplish this. There are a few challenges to overcome when we start to think about how our agent will perform these searches.

We currently have our embeddings stored in our vector store, so we'll need to create a retriever that will accept queries to search through these embeddings, such as *"Can you suggest hotels with good food?"*

Remember that we've stored our text embeddings with their own metadata. What would be even more useful is if we could also filter our embeddings by their metadata as you would with a relational database. For example, `hotel countryName` and `HotelRating` are both represented in our metadata. So, if we want to look for similar documents for hotels in a specific country with a specific rating, our question could be something like *"Can you suggest a four-star hotel in the UK with good food?"*

Luckily, Chroma DB provides the ability to filter queries on metadata, so we know we can achieve this with our vector store.

The next thing to consider is how we'll achieve this in our LangChain tool. The first challenge is that we want to extract parameters from our question so that we can use these downstream, specifically in our chroma query.

One approach could be to interact with an LLM to extract relevant search parameters from user interactions, which are then fed into your retriever to carry out a search against the Chroma DB. However, before diving in, it's always best practice to see what is already available in the LangChain ecosystem. As it happens, there is already something we can use, `SelfQueryRetriever`, a retriever built specifically to carry out this task. Let's break this down into detailed steps, explaining each component's role and how they integrate.

1. **Create the search service**: First, create a new file, `hotel_search.py`.

2. **Initialize OpenAI Embeddings**: Next, we initialize an embeddings model using the `OpenAI Embeddings`. This model is responsible for converting text into vector representations, which are used for semantic searches, and it's the same model we used to create our embeddings:

    ```
    from langchain_openai import OpenAIEmbeddings
    embeddings_model = OpenAIEmbeddings(
        model="text-embedding-3-small")
    ```

3. **Create a Chroma document store**: Next, we set up a Chroma document store that uses the embeddings from `embeddings_model` to perform semantic searches. Documents stored in Chroma can be retrieved based on their semantic relevance to query strings:

    ```
    from langchain.vectorstores.chroma import Chroma
    chroma_document_store = Chroma(
        collection_name="chapter8db_hotel_store",
        embedding_function=embeddings_model,
        persist_directory="chapter8db",
    )
    ```

4. **Define the metadata for the documents**: We define metadata attributes for the documents stored in Chroma. This metadata field info is going to be passed onto the retriever. For our example, we're only interested in searching by `HotelRating`, `cityName`, and `CountryName`. Feel free to add any other parameters you want to search by:

    ```
    from langchain.chains.query_constructor.base import AttributeInfo
    metadata_field_info = [
        AttributeInfo(name="HotelRating",
            description="Hotel rating", type="string"),
        AttributeInfo(name="cityName",
            description="Name of the city", type="string"),
        AttributeInfo(name="countryName",
            description="Name of the country", type="string"),
    ]
    ```

5. **Initialize the self-query retriever**: `SelfQueryRetriever` is initialized with the language model, document store, and metadata information that we defined in the previous step. It's configured to perform similarity-based searches within its own indexed content, finding documents that are semantically relevant to user queries:

    ```
    from langchain_openai import OpenAI
    from langchain.retrievers.self_query.base import SelfQueryRetriever

    # Initialize the language model with a specific temperature
    ```

```
llm = OpenAI(temperature=0.0)
retriever = SelfQueryRetriever.from_llm(
    llm=llm,
    document_store=chroma_document_store,
    document_content_description="Reviews and Descriptions of
hotels",
    metadata_field_info=metadata_field_info,
    verbose=True,
    search_type='similarity',
    search_kwargs={'k': 10}
)
```

We're also limiting the number of documents to return to 10 and setting verbose=True so that we can view the output. Let's move on to creating our prompt.

6. **Configure a prompt template**: Next, PromptTemplate is defined to ensure that the retrieved information from the documents is presented clearly to the LLM when answering user queries:

```
from langchain_core.prompts import PromptTemplate
template = """you can use the info below to answer questions.
Helpful Answer: HotelName: {HotelName} CountryName:
{countryName} CityName: {cityName} Address: {Address}
HotelRating: {HotelRating} HotelWebsite: {HotelWebsiteUrl}\n\
n{page_content}"""
prompt = PromptTemplate(
    input_variables=["HotelName", "HotelRating",
        "countryName","cityName","Address", "HotelWebsiteUrl"],
    template=template,
)
```

Now that we've created our prompt, we can bring everything together to create our tool:

7. **Create a retriever tool**: First, we wrap the retriever in a retriever tool using create_retriever_tool, passing in the title of the tool, the description to the retriever itself, and the document prompt we created:

```
retriever_description = """
"Searches and provides hotel suggestions by searching by
question, return the Hotel name, Hotel Rating, Hotel City,
Description, Website and a short summary from the metadata.
If there are no hotels matching, state that it can't find any
hotels with this criteria
"""

hotel_search = create_retriever_tool(
    retriever,
    "search_hotel_suggestions",
```

```
        retriever_description,
        document_prompt=prompt
    )
```

`create_retriever_tool` returns our tool of type `Tool`. The final step involves integrating `hotel_search` into our LangChain conversational agent, which we'll do by passing it on to our `AgentExecutor` when we create our agent. To recap, this agent will process user queries about hotel recommendations, leveraging the semantic search capabilities of `SelfQueryRetriever`, which now understands which metadata to look for and work with in our hotel questions, providing users with helpful hotel recommendations.

Weather information

We can create a custom tool to provide our weather information. The simplest way to create a custom tool is to annotate a function with `@tool`.

So, for our weather tool, it's straightforward to construct. We create our function named `search_weather`, which the decorator uses as the tool name, and the description is pulled from the docstring. For the weather data, we're going to use `OpenWeatherMap`, and to make the process of calling this API nice and simple, we'll use a Langchain utility, `OpenWeatherMapAPIWrapper`. Once again, it's worth pointing out that if you've found a service you want to consume for your ChatGPT app, always have a look around the LangChain ecosystem before building anything bespoke. Create a file named `search_weather.py` and add the following:

```python
from langchain.tools import BaseTool, StructuredTool, tool
from langchain_community.utilities import OpenWeatherMapAPIWrapper
os.environ["OPENWEATHERMAP_API_KEY"] = st.secrets.OPENAI_API_KEY
@tool
def search_weather(query: str) -> str:
    """A weather tool optimized for comprehensive up to date weather
information.
    Useful for when you need to answer questions about the weather,
use this tool to answer questions about the weather for a specific
location.
    Look in context for the location to provide weather for"""
    weather = OpenWeatherMapAPIWrapper()
    weather_data = weather.run(query)
    return weather_data
```

Go to `https://openweathermap.org/api/` and sign up for an account to get your API key. This will need to be available as an environment variable. We'll look at setting environment variables when we create the chatbot application.

Location information

We want to create a tool to answer questions about hotel locations with up-to-date data so that we can add these results to our context. There are a number of options. We could use the Tavily search engine, the purpose-built search engine for LLMs, and there is also an out-of-the-box tool in LangChain, `TavilySearchResults`.

However, a great place to find information is Wikipedia, so let's use this as the information source for our tool.

We're going to use an out-of-the-box LangChain tool to interact with Wikipedia, `WikipediaQueryRun`. You'll need to install the Wikipedia Python package by running `pip install --upgrade --quiet Wikipedia`.

It's easy to create this tool – create a file named `wikipedia.py` and add the following to create the Wikipedia API wrapper:

```
from langchain.tools import WikipediaQueryRun
from langchain_community.utilities import WikipediaAPIWrapper
api_wrapper = WikipediaAPIWrapper(top_k_results=5,
    doc_content_chars_max=500)
```

The wrapper provides some useful ways to fine-tune Wikipedia search results, so we've limited the number of results returned to 5 and limited the content for each returned document to 500 characters.

Next, create the tool by passing in our wrapper and giving the tool a description:

```
wikipedia_search = WikipediaQueryRun(
    api_wrapper=api_wrapper,
    description="look up tourist information for locations")
```

We'll now be able to use this tool in our agent. In the next section, we'll look at creating our LangChain agent using the tools we've created. In the next section, we're going to bring all these elements together by creating an agent that can utilize the tools we've created.

Bringing it all together – building Your own LangChain Chatbot with Streamlit

Let's begin constructing our web-based chatbot so that we can interact with ChatGPT using LangChain. To accomplish this, we'll use the Streamlit Python library to build the interface. Streamlit's wealth of ready-to-use components and utilities, as well as great documentation, allows us to swiftly develop a functional web-based chatbot. Its straightforward yet powerful features support quick prototyping and iterative refinement, making it perfectly suited for our chatbot project. Streamlit simplifies the integration of interactive elements and provides a lot of out-of-the-box functionality, which will cut our development time and allow us to concentrate on the LangChain project.

Make sure you've installed Streamlit with `pip install streamlit`, or by using one of the many other ways to install. I'm using Anaconda, so I'm using this distribution. You can test that you've installed Streamlit correctly by running `streamlit hello`.

In our project folder, create a folder named `app`, where we'll store our application code. Copy over the tools you created earlier into a tools folder under `app/tools`. Next, let's look at managing our API keys.

Creating secrets and config management

The first bit of Streamlit magic will be to use the out-of-the-box secrets manager. Streamlit provides native file-based secrets management so that we can easily store and securely access our API keys and manage projects. Create a folder named `.streamlit` and a file named `secrets.toml`, and then add the API keys we need for this project:

- `OPENAI_API_KEY="YOUR KEY"`
- `LANGCHAIN_API_KEY="YOUR KEY"`
- `OPENWEATHERMAP_API_KEY="YOUR KEY"`

Streamlit automatically loads secrets from `secrets.toml` into your project directory during development, and it provides options if you want to deploy via the Streamlit Cloud interface or make your application publicly available. Let's move on to creating our agent code.

Creating our agent service

Let's organize the code for our LangChain service in a dedicated LangChain service file.

1. **Create the LangChain service**: Begin by creating a Python file dedicated to your LangChain service and storing it in the app folder. This could be named something like like langchain_service. py. This file will contain all the necessary code to set up and run your agent.

2. **Import the necessary libraries and modules**: Import the os module for environment variable management, components from LangChain to access the hub functionalities and agent execution, our `ChatOpenAI` model, `AgentExecutor` to manage prompts and conversation memory, and the tools we created in the earlier steps (`weather_search`, `wikipedia_search`, and `hotel_search`), designed for specific chatbot functionalities.

3. **Define the** `setup_agent` **function**: This function is designed to configure and return an `AgentExecutor` instance, equipped with specific tools and settings for the chatbot:

```
def setup_agent(msgs: list, openai_api_key: str,
    langchain_api_key: str, openweathermap_api_key: str
) -> AgentExecutor:
os.environ["OPENAI_API_KEY"] = openai_api_key
```

```
os.environ['LANGCHAIN_TRACING_V2'] = "true"
os.environ['LANGCHAIN_ENDPOINT'] = "https://api.smith.langchain.
com"
os.environ['LANGCHAIN_PROJECT'] = "ellie_chatbot"
```

The preceding parameters include a list of messages for conversation history from the Streamlit app and the API keys for the OpenAI, LangChain, and `OpenWeatherMap` services. We then set our environment variables so that LangChain can access them. I'm also setting the environment variables so that we can track everything in LangSmith, along with a project name so that we can easily track all our interactions with Ellie.

4. **Initialize the language model, tools, and prompt**: An instance of `ChatOpenAI` is created with specific temperature and model parameters. We also create our list of tools:

```
llm = ChatOpenAI(temperature=0.0, model='
    gpt-4-1106-preview',verbose=True)
tools = [hotel_search,weather_search,wikipedia]
hub_prompt = hub.pull("hwchase17/openai-tools-agent")
```

The predefined tools (`hotel_search`, `weather_search`, and `wikipedia`) will add our specialized functionalities, which the agent will decide to use depending on the question asked. `hub_prompt` is retrieved using `hub.pull`, which fetches a pre-defined prompt template from the LangChain hub. You can create this manually if you want more control, but for our example, this works perfectly.

5. **Configure the conversation memory**: We are going to use `ConversationBufferWindowMemory`, which will give our Chatbot memory of the conversations and allow it to provide context-aware responses. Note that we're keeping the last 20 interactions in memory by setting `k=20`, and we are also passing in a list of messages (`msgs`) into `chat_memory` from our Streamlit agent, which allows us to tie up our agent memory with Streamlit Session State.

6. **Create the agent with the tools, memory, and the AgentExecutor**: Next, we create our agent with the imported `create_openai_tools_agent` function. This LangChain function creates an agent capable of using OpenAI's `tools` functionality:

```
agent = create_openai_tools_agent(llm, tools, hub_prompt)
```

`create_openai_tools_agent` is called with the language model (`llm`), the list of tools, and `hub_prompt` to create a configured agent.

7. **Create our Agent Executor**: Finally, we can create our `AgentExecutor`, which is instantiated with the created agent, tools, memory configuration, and verbosity settings. This executor manages the execution flow of the chatbot, processes inputs, and generatesresponses based on the integrated tools and conversation memory. The `setup_agent()` function returns the `AgentExecutor` instance, which is now ready to be used in the Streamlit chatbot application. We'll look at how we create this application and use our LangChain service in the next section:

Building our Streamlit chat app

With all our components now ready, it's time to build our chatbot. Start by creating a file named `main.py`; this file will serve as the primary Streamlit script. Within this script, we will develop Ellie's chatbot interface. Let's go through the steps to create this code:

1. **Import the libraries**: We need to import the essential libraries, including our LangChain history component and `streamlit (as st)` for our chatbot elements. We also import `setup_agent()`. This function from the `langchain_service.py` we created initializes the chatbot's agent with the necessary configurations and API keys.

2. **Initialize the Streamlit web application**: Define the `init_streamlit()` function; this sets up the Streamlit page configuration with `st.set_page_config`, specifying the page title and icon. We also give our page a title with `st.title("` 🦊 `Ellie the explorer ")`.

3. **Next, we initialize our chat message history**: This is important, as Streamlit apps operate on a single-threaded model, where the entire script reruns from top to bottom upon any user interaction that triggers a rerun. So, to save our messages, we would need to create a mechanism to manage these in Streamlit's Session State. We also need to consider handing these over to our LangChain app. To achieve this, we use a component from the LangChain Community library. `StreamlitChatMessageHistory` is a class that allows you to store and use chat message history in a Streamlit app,

   `msgs = StreamlitChatMessageHistory()`.

   It works seamlessly with LangChain by allowing you to incorporate the chat message history into your LangChain chains. It works by storing messages in Streamlit Session State at a specified key (the default is `langchain_messages`). We then pass over a reference of this memory to our agent.

4. **Define the `prepare_agent` function**: This is a helper function that calls `setup_agent` with the chat message history and API keys stored in `st.secrets`, returning a configured `AgentExecutor`.

5. **Create our `AgentExecutor` Langchain agent**: We call `prepare_agent()`. to create our `agent_executor`

   `agent_executor: AgentExecutor = prepare_agent(msgs)`, passing the chat message history and API keys from `st.secrets` to the `setup_agent()` function imported from our LangChain service code, `langchain_service.py`, which returns the `AgentExecutor` instance configured for the chatbot.

6. **Display a welcome message**: Add a welcome for the user at the start of the conversation, guiding them on how to interact with the chatbot:

```
if len(msgs.messages) == 0:
    msgs.add_ai_message(welcome)
```

The message is defined and added to the chat history if it's the initial load (i.e., there are no messages in the history).

7. **User interaction and response handling**: Next, we want to handle user input. If the user inputs a new prompt, we generate and draw a new response:

```
if prompt := st.chat_input():
    st.chat_message("human").write(prompt)
    try:
        response = agent_executor.invoke({"input": prompt})
        if 'output' in response:
            st.chat_message("ai").write(response['output'])
        else:
            st.error("Received an unexpected response format
from the agent.")
    except Exception as e:
        st.error(f"An error occurred: {str(e)}. Please try again
later.")
```

An optional component can be added so that you can easily look at what's being stored in `st.session_state.langchain_messages`. This is an expander widget (`st.expander`) to display the message history in the JSON format.

The application listens for user input through `st.chat_input()`. When input is received, it's displayed as a human message using `st.chat_message()`.

The input is then passed to `agent_executor.invoke()` to process the response, which, upon success, is displayed back to the user as an `ai` message.

We're also handling errors in the response format and errors coming back from our agent call; if there is an error, then this message is displayed using `st.error()`.

If the script is the main program (i.e., not imported as a module), `init_streamlit()` is called to start the app:

```
if __name__ == "__main__":
    init_streamlit()
```

That's it! Everything should now be in place to create a functioning chatbot. In the next section, let's take it for a spin.

Running and testing Ellie, our chatbot application

To run this Streamlit application, ensure that you have Streamlit and all the necessary packages installed in your Python environment. Then, navigate to the directory containing the `main.py` file and run the following command in your terminal:

```
streamlit run app/main.py
```

This command launches the Streamlit web server and opens the app in your default web browser, allowing you to interact with the chatbot. If there are any issues, double-check that your API keys and any other required configurations are correctly set in `st.secrets` and passed appropriately to the `setup_agent()` function for the application to function as expected. You should also be able to see your logs appearing in LangChain once you start interacting with the chatbot, enabling you to drill down into each run.

Let's try our first search for hotel recommendations. We get a useful result, as you can see in the following screenshot of the output:

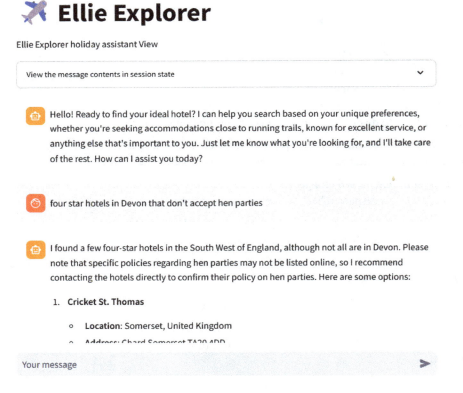

Figure 8.2 – Asking our agent for hotel recommendations

The agent looks at the question and chooses which tool to use to fulfill the request. This question is hotel-specific, so the agent calls our `hotel_search` tool to answer our question. Remember to look at the debug output in LangSmith to view the individual runs for the project. This will allow you to see the individual calls to get a solid understanding of how your agent interacts with the different LLMs to decide on the tool to use and return our response, as well as other tasks specific to the individual tool – for example, extracting the parameters to use in filtering our vector search. The following LangSmith screenshot shows this:

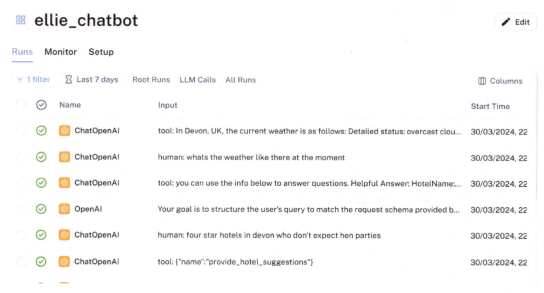

Figure 8.3 – View the calls from our agent to the LLMs in LangSmith

Feel free to try your own questions with anything specific to your own data.

We should also be able to ask follow-up questions so that the agent exercises the other tools. In the following screenshot, we also ask about the weather, and you can see Ellie responds with data from our weather search tool.

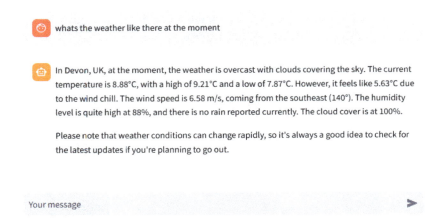

Figure 8.4 – We ask a weather-specific question to test our weather tool

Try out the `wikipedia_search` tool by asking for something specific about the area.

So, we now have a working ChatGPT-powered agent that can form a good basis for an even more advanced agent. Let's look at what other improvements you can carry out.

Ways to improve Ellie

There are lots of different enhancements you could implement to improve Ellie's travel assistant capabilities:

- Improve the chat UI – for example, implement streaming answers
- Provide more complex interactions by enhancing our LangChain agent so that we can carry out more complex transactional tasks
- Learn user preferences – their interests, family situation, and what they like to do on holiday
- Enhance the chatbot memory with a vector store to persist conversations so that they can be used between sessions
- Look at the chunking strategy to improve vector embeddings for better searches

So far, what you've created is just a start and a good basis for creating a more complex ChatGPT-powered agent. Any of these enhancements would be worth implementing to improve Ellie's capabilities.

Summary

In this chapter, we looked at a real-world example of creating a ChatGPT-powered conversational agent, using some of the techniques and technologies that we covered in earlier chapters. Our example was thoughtfully designed to cover the use cases, functionality, and features of our chatbot, allowing us to apply our knowledge of LangChain, RAG systems, and related tools effectively. You were also introduced to the Streamlit library to create rich chat interfaces and saw how to use this with LangChain. Hopefully, we've implemented our code base to make our agent extensible and something that we could use for a more complex project. We've achieved our aim of creating a ChatGPT conversational agent with knowledge of our own data and the capability to use external information and sources.

We've created our POC and validated our concept. It's a start, but there is a lot to consider before creating a more extensive application and going live. In our final chapter, we'll consider going into production with what we've learned so far and what other core functionality we would need to create to make this a reality. We'll also consider other alternatives to ChatGPT and the emergence and importance of smaller language models. Finally, we'll come to some conclusions and see where we go from here with the current state of the technology.

Further reading

The following links are a curated list of resources that were used in this chapter:

- Streamlit:

    `https://streamlit.io/`

- LangSmith:

    `https://smith.langchain.com/`

- Embedding models:

    `https://python.langchain.com/docs/integrations/text_embedding` and `https://platform.openai.com/docs/guides/embeddings/embedding-models`

- Chroma DB:

    `https://docs.trychroma.com/`

9

The Future of Conversational AI with LLMs

In this chapter, we dive deeper into topics related to taking our ChatGPT app to production. We've already touched on some of these areas throughout the book, but now let's consider some of these aspects in more detail as well as look at lessons already learned in the industry. Building on what we've learned from creating ChatGPT applications so far, we can make informed decisions and create successful strategies related to our conversational AI projects.

We'll also explore other alternative technologies to ChatGPT at the time of writing, with a special focus on the potential of smaller language models. Finally, we'll look ahead to what the future might hold for LLMs and where your organization may be heading, preparing you for what's coming next in the evolving landscape of conversational AI.

In this chapter, we'll cover the following topics:

- Going into production
- Evaluating production systems
- Learning how to use LangSmith to evaluate your project
- Advanced monitoring features with LangSmith
- Alternatives to ChatGPT and LangChain
- The growth of the **small language model** (**SLM**)
- Where to go from here

Technical requirements

In this chapter, we will be using ChatGPT extensively, so you will need to be signed up with a free account. If you haven't created an account, go to `https://openai.com/` and click **Get Started** at the top right of the page.

The examples requires python 3.10 and Jupyter notebook to be installed `https://jupyter.org/try-jupyter/notebooks/?path=notebooks/Intro.ipynb`. The examples require Python 3.10 and Jupyter Notebook to be installed, which you can do by going to `https://jupyter.org/try-jupyter/lab/index.html`.

You can find the chapter code here: `https://github.com/PacktPublishing/ChatGPT-for-Conversational-AI-and-Chatbots/chapter9`.

Going into production

You've made some great progress in increasing your knowledge of how to create a ChatGPT-powered conversational agent as you've progressed through the book. Hopefully, you've also been educated on some of the limitations, challenges, and pitfalls of creating systems powered by LLMs and, most importantly, getting them production-ready. These challenges are not to be underestimated, and it's no surprise that despite the huge adoption of LLMs over the past year or so, examples of conversational AI agents entirely powered by LLMs are thin on the ground, and the ones that are have come under some intense scrutiny and unfortunately have underperformed in some cases. Let's look at one in the next section.

Understanding the dangers of going into production

One recent example of how LLMs in production can quickly land you in hot water is the issues found with a chatbot deployed by a delivery firm of note. I'm not going to finger-point here by naming the company in question; there are other examples of misbehaving chatbots. Another example involved buying a Chevy Tahoe for a dollar – I suggest you Google that one!

For the delivery chatbot, one disgruntled customer decided to inject some humor into their conversation by asking a series of awkward questions, resulting in the chatbot swearing and delivering a poem calling out the delivery company as useless. Not the greatest conversation for a customer support chatbot to be having, but a great example of what can go wrong with any LLM-powered conversational agent without the appropriate guard rails.

So, what happened here? I can't comment on the tech stack or LLM technologies used by the company in this chatbot project. However, if an agent does start responding with swear words, it looks like they were not employing any output validators or content moderation checks. This allowed the member of the public to essentially prompt the LLM directly themselves. This looks unlikely to have been an issue with model training but more an issue caused by missing guard rails to stop malicious (OK – a bit of an exaggeration) prompting. This allowed the user to engage the LLM in off-brand topics or out-of-scope requests. So, content filtering on LLM questions would have stopped this, coupled with response checking to sanitize output for appropriate content before sending it to the client.

There are really two issues here: input and output, which you would hope would be handled by a more solid prompting strategy. For example, if the agent were a LangChain agent powered by ChatGPT, then we'd look at creating a strong system prompt and using an LLM call to check an initial question before we even entered our main pipeline. This would need to be backed up with an extensive evaluation process covering not only standard questions but also how the system handles users trying to manipulate and/or break the system so that they can post about it on social media.

Remember – with great power comes great responsibility, and when creating LLM projects, it's a different ballgame from traditional **natural language understanding** (NLU) systems. We've essentially gone from making a system and watching out for correct, incorrect, and missing answers to a system where all this can happen and practically anything else if you're not making a robust enough effort to stop it.

By now, I'm sure you appreciate that one of the most popular LLM applications is the RAG system, and we've covered this extensively in previous chapters. However, these are also not without their challenges; let's move on to some of these in more detail.

Challenges of RAG systems

As we've learned, RAG systems are one of the popular implementations in LLM conversational AI and may feature in systems you are currently trying to re-create with LLM technology or with the greenfield project you're developing. You've also learned you can get some pretty good performance out of these systems without huge amounts of effort. However, despite being the poster boy of LLM-powered conversational AI, the RAG system does pose challenges that become much more important as you move toward production, and we'll look at these challenges in the next section. So, let's consider we're looking at bringing a project with a RAG system into production – for example, the project in *Chapter 8*: our **proof of concept** (POC) for our travel assistant. Key stakeholders have been impressed by Ellie Explorer so far and want to release her as a web-based chatbot along with broadening her scope to provide more targeted personalized hotel suggestions and to answer FAQ questions and more transactional use cases. To recap, we used a RAG system so that we could inject contextual reference data from our LLM-ready datastore of hotel information to provide succinct, believable hotel recommendations.

Despite their innovative approach, naive RAG systems have several challenges that can impede their effectiveness and reliability.

It's a big step to move from a POC to production, and we'll need to consider RAG system challenges before releasing Ellie into the wild.

Traditional RAG shortcomings

Let's look at some of the shortcomings of RAG systems and how these may manifest in our systems in more detail:

- **Missing content**: Failures can occur when trying to handle a question that cannot be answered from the existing underlying document data, maybe because the information is just not there or there is some issue with the chunking or embedding process. This can be hard to debug and needs to be handled correctly. Ideally, the RAG system will simply reply with a message such as "Sorry, I don't have this information." However, in cases where questions are relevant to the content but lack specific answers, the system might be misled into providing a response. Or an even harder failure to spot can be taken from our previous example: if no results are returned from the RAG system search – for example, hotel recommendations for a country not in our database of provided hotels. The LLM was responding itself from its own knowledge of hotels, not those from the RAG results, so suggesting hotels outside of the company hotel listings. This behavior can be rectified with prompting, but you can see the issue here: it's hard to spot. Realistically, we need to check our hotel results suggestions intermittently to ensure this does not occur.

- **Inadequate retrieval and precision issues**: A crucial challenge for RAG systems is the retrieval of relevant information. The underlying retrieval model may fetch pertinent data, which may result in the generation of text that is misinformed or irrelevant. This often occurs due to training on data of insufficient quality or data that does not accurately represent the intended domain. Enhancing data quality and ensuring its alignment with the target domain are critical steps toward addressing this challenge. However, this may also show we need more sophisticated retrieval algorithms that can accurately understand the context and nuances of queries.

  There can also be issues with results ranking, where the document contains the answer to the question but didn't rank high enough to be presented to the user. There can also be issues with results ranking, where the document contains the answer to the question but it didn't rank high enough to be presented to the user. In the case of Ellie, if we are returning n hotel results, in theory, all documents are ranked, but this may mean some don't make the cut to be sent over to the context.

  Basically, if we are injecting prompts with incorrect or poor contextual reference then it's no surprise that the resulting response may not provide good results. Basically, if we are injecting prompts with incorrect or poor contextual reference, then it's no surprise that the resulting response may not provide good results.

- **Outdated information**: RAG-generated content may be based on outdated facts if the knowledge source isn't regularly updated or if the model lacks mechanisms to discern and incorporate temporal information. Keeping the knowledge base current and designing retrieval systems to account for the time-sensitivity of information is important.

- **Increased RAG costs**: There are concerns that there is going to be a growing unnecessary overhead of token costs with many ChatGPT systems with a RAG component, primarily caused by unoptimized retrievals and additional text results being passed into context.

- **Is a RAG system too dumb?**: In all honesty, so far, what we have covered is a moderately straightforward RAG implementation. It's obvious that a RAG system needs to be considering a lot, and the latest RAG systems are making more demands, with three core themes central to these systems that need to perform well if you are going to production: .

  - Knowing when and where to retrieve data when satisfying a question. There can potentially be multiple vector stores to pull data from.

  - Evaluating the data from your retrieval by quality checking or ranking results. Using an LLM to vet your output is potentially a solution but can result in errors.

- You may need to carry out multiple retrieval calls and then select the best result. Practical evaluation of results is hard, and verifying multiple facts independently can be time-consuming and affect system efficiency.

In our Ellie example, our `AgentExecutor` will be tackling *a*, but *b* and *c* are something you'll need to build into your system.

- **Common LLM issues**: Hallucination, Toxicity, Irrelevence

Hallucinations and irrelevance can arise from flaws in the system design or inaccuracies within the knowledge base as well as toxicity and bias from the underlying RAG data or surfacing through response generation from the ChatGPT call. Hallucinations and irrelevance can arise from flaws in the system design or inaccuracies within the knowledge base, as well as toxicity and bias from the underlying RAG data or surfacing through response generation from the ChatGPT call. All are issues that need to be watched for in a production system.

So, now we've covered the challenges and solutions you're going to face, in the next section, let's consider how we can evaluate your LLM application.

Evaluating production systems

Developing a prototype ChatGPT application with a RAG component is relatively straightforward, but preparing it for production and maintaining it effectively calls for consistent evaluation, which is challenging. Using LLMs to assist in evaluation is an obvious approach; however, be aware that there are pitfalls caused by LLMs displaying positional bias, answer-style preference, and variable results from one run to the next, which leads to the need for a more structured approach. It's also worth remembering that the use case and nature of an agent will influence your evaluation approach. For example, a simple RAG question and answer compared to a more conversational agent will need a different method of evaluation.

It's important to understand that, as in any **machine learning** (**ML**) project, you should evaluate the RAG pipeline's performance using a validation dataset and an evaluation metric. This involves assessing the components individually and collectively: knowledge retrieval, response generation, and their integration. There are a number of different evaluation metrics you can consider, such as context relevance, groundedness, and answer relevance.

Currently, determining the right evaluation metrics and collecting good validation data is a quickly evolving topic. In an attempt to tackle drawbacks with LLM-assisted evaluation, there are various methods and metrics for assessing RAG systems, including tools such as **Retrieval Augmented Generation Assessment** (**RAGAs**) or the RAG Triad of metrics, along with a range of tooling options and libraries. Each one of these merits more than we can cover here, and it's worth reading about a few of them to see different approaches and look at how these can be implemented. These are primarily methodologies designed to evaluate the effectiveness of RAG applications in a structured and comprehensive manner by focusing on the precision of knowledge retrieval and the quality of response generation. Let's consider the process of evaluation for a LangChain application using LangSmith in the next section

Components of an evaluation system

LangSmith offers a robust evaluation framework designed to monitor and improve LLM applications through well-structured components outlined in the following diagram:

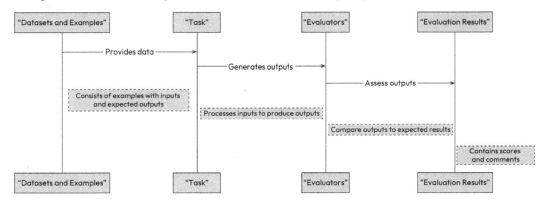

Figure 9.1 – Process of evaluation

LangSmith offers a UI and SDK for building datasets and editing and versioning them: an SDK for defining your own evaluators or creating and using custom evaluators and a UI for carrying out inspections and trace analysis.

Let's go through the components of an effective evaluation system in more detail:

- **Datasets and examples**:

    The fundamental unit of any evaluation system in LangSmith is the dataset, which is comprised of multiple examples. Each example within a dataset represents a test case that contains both inputs and expected outputs:

    - **Inputs**: These are the data points that the LLM will process.

    - **Outputs**: These are the expected results that the LLM should produce when given the inputs.

    Datasets are typically curated from various sources, including manual creation, user feedback, or conversation examples that have gone well or generated outputs from LLMs. If you're looking at migrating to a ChatGPT-powered agent from an existing system, then your historical conversation data is going to be vital for evaluation.

- **Types of datasets**:   There are different types of datasets supported by LangSmith:

    - **KV**: Key-value pairs that accommodate multiple inputs and outputs.

    - **LLM**: Direct string inputs and outputs, typically used in simple prompt-response models.

    - **Chat**: Structured like chat conversations, suitable for models trained on dialog.

It's easy to create datasets in LangSmith, where you can create and edit datasets and example rows as well as import from CSV.

- **Task**: The task is the actual operation or model being evaluated. It processes the inputs from the examples in the dataset and generates outputs, which are then assessed by the evaluators.

- **Evaluators**: Evaluators are functions or mechanisms that assess the outputs generated by the task. They compare these outputs against the expected results or apply specific criteria to determine the quality of the outputs. Evaluators return scores and can provide feedback for each example. There are a number of different types of evaluators:

  - **Heuristic evaluators**: These are simple functions that perform checks such as validating output formats (for example, JSON validation) or string matching against expected outputs.

  - **LLM-as-judge**: Advanced evaluators that use another LLM to assess the quality of outputs based on content criteria such as offensiveness or relevance.

  - **Human**: Manual review of outputs, often used for final validation or in cases where nuanced judgment is required.

- **Evaluation results**: The output of an evaluation is typically an `EvaluationResult`, which contains the following:

  - **Key**: The metric name being evaluated (for example, accuracy, completeness).

  - **Score**: The quantitative measure of how well the output matched the expected result or met the evaluation criteria.

  - **Comment**: Optional insights or reasoning behind the score.

Now you've got an understanding of the components of an evaluation system, in the next section, we'll look at bringing this into practice with an example.

Learning how to use LangSmith to evaluate your project

Remember that different types of AI agents will require different types of evaluation approaches. There is no one-size-fits-all approach, and it's up to you to decide on the methodology and approach to follow based on your use case. A Q&A agent would be simpler to evaluate if you are looking to support knowledge of a specific domain – for example, company-specific information from a RAG system – while an agent that needs to support transactional conversations will need a more complex evaluation implementation as you need to be absolutely sure your conversational agent is going to consistently meet their task.

With an intent-based system, you are able to accurately control each step of the conversation, while with an LLM-powered conversational agent, you're controlling your agents' actions and capabilities with prompts, which in my opinion is a more nuanced and volatile approach.

Langsmith makes this kind of in-depth specific evaluation possible let's run through a simple example in code so you can see the basics working of working with LangChain:LangSmith makes this kind of in-depth specific evaluation possible. Let's run through a simple example in code so that you can see the basics of working with LangChain.

Our example is made up of a RAG system that answers questions about ultramarathon running from a simple knowledge base:

1. **Process source documents and index**: For our knowledge base, we're going to use a `source.text` text document source.text with some information about running. Feel free to use your own subject matter – I just like running! We follow a standard workflow outlined in earlier chapters to create a simple LangChain RAG based on a simple text corpus to evaluate – document chunking, embedding creation, and creation of a retriever with OpenAI embeddings and an in-memory Chroma DB for the vector store:

    ```
    loader = TextLoader('source.txt')
    documents = loader.load()
    text_splitter = RecursiveCharacterTextSplitter(
        chunk_size=500,
        chunk_overlap=20,
        length_function=len,
        is_separator_regex=False,)

    split_documents = text_splitter.split_documents(documents)
    embeddings = OpenAIEmbeddings(model="text-embedding-3-small")
    chroma_document_store = Chroma.from_documents(
        split_documents, embeddings)
    retriever = chroma_document_store.as_retriever(
        search_kwargs={"k": 4})
    ```

2. **Create a simple LangChain RAG to evaluate**: In the following code, we create a standard RAG pipeline to interact with our Chroma retriever. Our pipeline setup should be pretty familiar to you by now. Initialize an instance of the `ChatOpenAI` class and create a `ChatPromptTemplate` instance from a template string with placeholders for questions and context that will be filled out during the execution of the RAG pipeline. Next, bring everything together using **LangChain Expression Language** (**LCEL**) to create our chain, which you can invoke to check everything is working correctly:

    ```
    model = ChatOpenAI(model="gpt-3.5-turbo-0125")
    template = """You are a helpful documentation Q&A assistant,
    trained to answer questions about ultra marathon running.
    ```

```
Use the following pieces of retrieved context to answer the
question.
If you don't know the answer, just say that you don't know.
Use two sentences maximum and keep the answer concise.
Question: {question}
Context: {context}
Answer:
"""
prompt = ChatPromptTemplate.from_template(template)
# Setup RAG pipeline
rag_chain = (
    {"context": retriever,  "question": RunnablePassthrough()}
    | prompt
    | model
    | StrOutputParser()
)
chain.invoke("do i need to do any long runs in my training")
```

3. **Install LangSmith and initialize your client**:

    Make sure you have installed LangSmith by running `pip install -U langsmith`. Then, set up your environment to use LangSmith, then create your LangSmith client:

    ```
    from langsmith import Client, evaluate
    from langsmith.schemas import Run, Example
    # Initialize the LangSmith client
    client = Client()
    ```

4. **Create and populate a dataset**:

    For QA correctness evaluation, create a dataset that specifically includes questions about your specific knowledge base. Each example should consist of a question and the correct answer. We can do this by importing from CSV or programmatically: we create a name and description then call client.create_dataset() to create the dataset, then add our examples with create_examples(). We can do this by importing from CSV or programmatically: we create a name and description, then call `client.create_dataset()` to create the dataset, then add our examples with `create_examples()`. We can also add these directly using the LangSmith UI:

    ```
    dataset_name = "Ultra Marathon QA Dataset"
    description = "Dataset for evaluating QA correctness of an ultra
    marathon chatbot."
    dataset = client.create_dataset(dataset_name,
        description=description)
    client.create_examples(
        inputs=[
    ```

```
        {"question": "How do...", "context": "..."},
        {"question": "will i...", "context": "..."}
    ],
    outputs=[
        {"answer": "Training..."},
        {"answer": "Consider buying..."}
    ],
    dataset_id=dataset.id,
)
```

Another easy option to add examples to datasets is from existing runs in LangChain. Just go to the **Run Details** page and click the **Add to Dataset** button.

5. **Define an evaluation method**:

Create a function to call the thing we are evaluating, in our case it's our RAG chain, but this could be any runnable element of your pipelineCreate a function to call the thing we are evaluating. In our case, it's our RAG chain, but this could be any runnable element of your pipeline:

```
def predict(inputs: dict):
    return rag_chain.invoke({"question": inputs["question"]})
```

6. **Define an input formatter for the evaluator**:

Define a function to format inputs for the evaluator. This step ensures that the evaluator function receives the necessary data in a consistent format:

```
def format_evaluator_inputs(run: Run, example: Example):
    return {
        "input": example.inputs["question"],
        "prediction": next(iter(run.outputs.values())),
        "reference": example.outputs["answer"],
    }
```

7. **Configured evaluator**:

In this step, we'll set up the LangChainStringEvaluator. In this step, we'll set up the LangChainStringEvaluator evaluator. This involves configuring it to assess the correctness of the chatbot's responses based on predefined criteria and normalization factors. This evaluator will handle string comparisons, allowing for a precise measurement of answer accuracy:

```
from langsmith.evaluation import LangChainStringEvaluator
correctness_evaluator = LangChainStringEvaluator(
    "labeled_score_string",
    config={"criteria": "correctness", "normalize_by": 10},
    prepare_data=format_evaluator_inputs)
```

Let's look at the details of creating the evaluator:

- Label: 'labeled_score_string' specifies the scoring label for output in the evaluation results.**Label**: `"labeled_score_string"` specifies the scoring label for output in the evaluation results.

- **Config**: The configuration includes the following:

  - "criteria": "correctness" which defines what the evaluation should measure—here, the accuracy of the responses. `"criteria": "correctness"`, which defines what the evaluation should measure—here, the accuracy of the responses.

  - `"normalize_by": 10`, which adjusts the score to a scale of 0 to 10, making it easier to interpret the results.

- Prepare Data: format_evaluator_inputs is a function that formats the inputs, predictions, and references in a way that the evaluator can process effectively.**Prepare data**: `format_ evaluator_inputs` is a function that formats inputs,   predictions, and references in a way that the evaluator can process effectively.

8. **Run the evaluation with the configured evaluator**:

    Once the evaluator is configured, the next step is to execute the evaluation. This involves running the prediction function across the dataset and using the evaluator to assess each response.

    Notice we are using `evaluate()` and not `run_on_dataset()`, which is being deprecated but will still return evaluation results:

```
results = evaluate(
    predict,
    data=dataset_name,
    experiment_prefix="Chat Single Turn",
    evaluators=[correctness_evaluator],
    metadata={"model": "gpt-3.5-turbo"},
)
```

Let's look at the different elements of the `evaluate()` call:

- **Prediction**: The prediction function (`predict`) is what the chatbot uses to generate answers from the inputs.

- **Data**: Specifies which dataset to use for the evaluation. This dataset contains the questions and correct answers.

- **Experiment prefix**: This string helps to categorize and identify the evaluation results within the system.

- **Evaluators**: The list includes the `LangChainStringEvaluator` evaluator, set up in the previous step. You could also create your own custom evaluator and include it here; you can use multiple evaluators.

- **Metadata**: Contains details about the model version used, facilitating further analysis and traceability.

Keep in mind that LangSmith will also create evaluation code to run when you choose to create an experiment from the UI. This also provides examples of different pre-built evaluation chains, allowing you to evaluate for the following:

- Correctness
- Conciseness
- Relevance
- Coherence
- Harmfulness
- Maliciousness
- Helpfulness
- Controversiality
- Misogyny
- Criminality
- Insensitivity

At the time of writing, the code examples use `run_on_dataset()`, but it would be fairly straightforward to migrate this to `evaluate()`. When you run `evaluate()`, this should create an experiment in LangSmith, and you should see something such as the following as output:

View the evaluation results for experiment: 'Chat Single Turn-46c45783' at:

```
https://smith.langchain.com/...
```

The link should take you through to an experiment link in LangSmith where you can see details of each experiment, outlined in the following screenshot:

Figure 9.2 – Evaluation output shown in the experiment view in LangSmith

Clicking on the **SUCCESS** or **FAILURE** button will display details of the test for each input and output in the dataset. From this view, you can also drill down into each run to see what happened and via other information, such as the token costs for each run. Token costs for evaluation could quickly rise depending on how large your dataset is.

Let's try to test the prompt we've employed to stop the LLM from using other data and feeding back to the user if it can't answer. Try adding another question-answer pair to your dataset in LangSmith, but this time, let's add one that it shouldn't be able to answer. In my example, it's something specific about trainers that I know is not mentioned in my knowledge base:

```
Input: {"question": "Do I have to use carbon inserts in my trainers"}
```

output: {"answer": "The documents provided do not contain information about using carbon inserts in trainers for ultra marathon running."}

When I rerun the evaluation, we can see the test passes, and our agent is responding how it should. You can see the output in the following screenshot:

Figure 9.3 – Evaluation output displaying test for hallucinating in LangSmith

Try looking at other types of out-of-the-box evaluators to see if you can evaluate for other criteria.

There is no one-size-fits-all approach for evaluating different LLM-powered systems. Different types of conversational AI agents will require different types of evaluation approaches and different datasets. Look on the LangSmith website for examples of testing different types of agents.

The aim of any evaluation system is to offer a robust framework for evaluating your agent. LangSmith makes it easy to run evaluations and provide detailed insights into both individual and integrated component performance. With LangSmith, you can better prepare your ChatGPT application for the demands of a production environment, ensuring it delivers high-quality, reliable responses.

Remember – the key to successful AI development is continuous testing and refinement. Evaluations are not a one-time task but a regular part of the development cycle. In the next section, we're going to consider application monitoring.

Application monitoring in production with LangSmith

Monitoring any LLM-powered application is hard, and until the release of LangSmith, it's been difficult to have full exposure to what is going on under the hood in LangChain apps. Detailed monitoring of your application is vital to keep things running smoothly, and LangSmith offers numerous features to do this.

Advanced monitoring features with LangSmith

LangSmith's monitoring system enables effective oversight of production applications by allowing developers to configure and manage tracing. Key to this system are features tailored to optimize the collection and use of data during production.

Tracing and data management

LangSmith's tracing and management capabilities provide selective data point logging through sampling, improved data organization with metadata, and critical insights via feedback integration:

- **Sampling**: Essential in production, this feature allows for the logging of only a subset of data points to manage volume and relevance effectively

- **Metadata**: Attaching metadata to runs enhances the ability to filter and group data, facilitating more targeted analysis

- **Feedback**: Integrating user feedback helps highlight significant data points, drawing attention where it's most needed

Monitoring tools

LangSmith provides a comprehensive suite of monitoring tools designed to help with monitoring your application. Here are some key features:

- **Filtering functionality**: Users can delve into specific runs using advanced filtering options, which include filtering by name, metadata, feedback, and full-text searches.

- **Monitoring charts**: The **Monitoring** tab on the **Project** dashboard provides visual representations of various metrics such as **Trace Latency**, **Tokens per Second**, and **Cost**. These charts support temporal analysis and can drill down into specific data points for detailed trace tables, aiding in debugging.

- **Metadata grouping**: This allows for the comparison of different application versions side by side within the same charts, which is particularly useful for A/B testing changes.

Advanced monitoring features

LangSmith offers other advanced features to help you understand complex conversations and automate some tasks.

- **Threads**: For applications where traces are part of the same conversation, LangSmith offers threading by metadata keys to group related traces conveniently.

- **Automations**: To reduce manual oversight, LangSmith automations can be configured to handle repetitive tasks. After setting up filters, developers can automate actions such as sending data to datasets, annotation queues, or conducting online evaluations.

A fully featured monitoring tool was lacking in the LangChain space, and LangSmith fills this gap.

It stands out as an essential tool for you to ensure the robust performance of your LLM application. By providing detailed metrics and real-time data visualization, LangSmith should allow you to maintain high standards of performance, reliability, and user satisfaction in your conversational AI solutions.

Alternatives to ChatGPT and LangChain

Since the early days of ChatGPT, there have not only been huge advancements in LLM technology but also the availability and choice of service offerings. In this section, we'll look into the diverse range of service offerings that have emerged, exploring some viable alternatives beyond ChatGPT and LangChain.

Some alternatives to ChatGPT and OpenAI LLMs

Early last year, you could have argued that OpenAI's **Generative Pre-trained Transformer** (**GPT**) LLMs were in a one-horse race and would likely be your first choice in building an LLM-powered project. Nowadays, there are many companies vying to release the best model, resulting in a growing number of LLMs to choose from, both open source and closed, to create a conversational AI system. The LLM arms race continues; let's look at some examples, and it's worth noting this is not an exhaustive list:

- **Meta**: The Llama model family are popular LLMs provided by Meta. The latest models, Llama 3 models with 8B and 70B parameters, have shown impressive performance when tested against other LLMs and are available in several ways.

- **Anthropic**: Anthropic is known for its focus on building reliable and interpretable AI models. Their main model, Claude, is designed to enhance safety and controllability, aiming to reduce harmful or misleading outputs. Anthropic emphasizes ethical AI development, leveraging techniques such as **reinforcement learning from human feedback** (**RLHF**) to fine-tune model responses, ensuring they align closely with human values and safety expectations.

- **AI21**: AI21 provides access to the Jurassic-2 and Jamba transformer models. Jurassic-1 is AI21 Labs' flagship family of models. Jurassic-1 comes in various sizes, ranging from smaller models suitable for cost-effective applications to very large models that can generate more complex and nuanced text. These models are designed to perform a wide range of tasks, from answering questions and summarizing text to generating content and code.

- **Google**: Google never misses a chance to remind us they invented the transformer architecture and have created strong models such as PaLM 2 and, most recently, the Gemini series. The Gemini series, particularly the latest Gemini 1.5, pushes boundaries with a huge context window of 1 million tokens and allegedly better performance. The 1.5 series is also available in 3 sizes: Ultra, Pro, and Nano. Google has also renamed Bard to Gemini. So far, the Gemini performance I've seen is on par with GPT-4 models.

- Cohere: Cohere focuses on natural language understanding and generation. **Cohere:** Cohere focuses on NLU and **natural language generation (NLG)**. Their models are designed for a variety of applications such as content generation, summarization, and semantic search. Cohere's models are noted for their user-friendly API and robust security features, suitable for enterprise environments where data privacy and application integrity are critical.

- **MistralAI:** MistralAI positions itself as an enterprise-focused LLM with a privacy-preserving infrastructure. They offer open source models: Mistral 7B, Mixtral 8x7B, and Mixtral 8x22B, which are customizable. They can be downloaded or used on demand on the Mistral platform. The company also offers a large and small optimized commercial model.

- **Cerebras:** Cerebras released a family of 7 GPT models ranging from 111 million to 13 billion parameters. They claim to be some of the fastest and cheapest models to train. Cerebras also provides cloud infrastructure for training models and Cerebras AI Model Studio, which is a purpose-built platform for training and fine-tuning LLMs.

A great way to test some of these models side by side is by using Hugging Face Chat where you can choose from the popular LLMs to answer your question. You'll notice that all LLMs have a subtle nuanced style to their responses unless prompted to do otherwise. This observation came out of a conversation I had with the great Kane Simms from VUX World.

There is evidence that the gap between LLMs is narrowing. Recent performance comparisons suggest that the gap between different LLMs is narrowing. Models such as Claude 3, GPT-4, and Gemini Ultra are achieving similar scores in tasks such as multi-choice questions and reasoning. Interestingly, some smaller models such as Mixtral and Llama2 are showing better performance than larger models in certain areas – for example, reasoning – highlighting that it's not always the case that the largest models are the best.

We've been using LangChain a lot in this book. Let's look at some alternatives to LangChain in the next section.

Some alternatives to LangChain

I'm a big fan of LangChain, and now it's fully supported by LangSmith for monitoring and running your application in production, it's a solid platform. However, LangChain is not for everyone, with many defining factors in play before making a decision. There are many other options. You could build your own application in code and interact with an LLM API directly and still use LangSmith for monitoring. You could also use something such as **Generative AI Large Language Model for Enterprise (GALE)**: the new productivity suite by Kore.ai, which is a no-code solution to use the latest LLM models.

Alternatively, there are options such as LlamaIndex. LlamaIndex is a comprehensive data framework tailored to enhance applications built on LLMs.

It supports a large range of data formats, such as APIs, PDFs, documents, and SQL databases, allowing you to ingest, organize, and structure your data. The platform provides an advanced retrieval interface and the ability to create prompt chains to create LLM applications and evaluation capabilities, allowing you to create advanced RAG systems.

LlamaIndex, with its supporting Python and TypeScript SDKs and a vibrant community offering a wealth of resources, such as connectors, tools, and datasets, exemplifies the innovative solutions emerging in the LLM field. This growth in LLM-specific services marks a significant trend in the AI landscape, which we'll explore further in the following section.

Looking at the growing LLM landscape

Since ChatGPT was released last year, the LLM landscape has changed dramatically in the world of LLMs, and it's important to have an understanding here to help with any technology decision you may be making for your ChatGPT or LLM project.

There are more and more ways to access LLMs, which can be primarily broken down into the four methods illustrated in the following diagram:

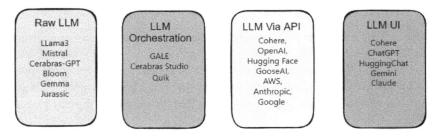

Figure 9.4 – Different ways to interact with LLMs

Let's look at the pros and cons of the four different ways to access and utilize LLMs:

- **Open source raw LLMs**: For those with technical expertise, using open source models on your own infrastructure offers a customizable solution and means to fine-tune your own specialist models, albeit with potentially higher upfront technical challenges and huge infrastructure costs. There is also a growing landscape of open source tools to run these models; for example, the llama.cpp library. Although free to use, the costs related to hosting and operating these models can be significant, requiring specialized knowledge. These expenses and the need for ongoing maintenance tend to increase as usage grows.

- **LLM orchestration**: Platforms that provide specialist cloud infrastructure for model hosting and training while providing an easy way to orchestrate and expose these models for use and the means to fine-tune them for specific domain expertise and use cases. Most of the **cloud service providers (CSPs)**, such as Google Cloud, **Amazon Web Services (AWS)**, and Microsoft Azure, provide **infrastructure as a service (IaaS)** to host your models. This can be expensive but offers the ability to tap into open source LLMs without the infrastructure costs. There are also some promising platforms which allow you to create end to end conversational assistants powered by LLMs for example Quiq.

- **LLMs via APIs**: This is the most common method used by organizations to integrate LLM capabilities into their applications. APIs provide a straightforward path to developing applications with LLMs. However, using these APIs comes with several challenges, such as cost implications, concerns over data privacy, issues with inference latency, rate limits, risks of catastrophic forgetting, and model drift.

- **LLM UI**: Not really needing any introduction, interfaces such as ChatGPT, HuggingChat, Cohere Coral, and Gemini provide conversational experiences directly to users and have become part of many people's day-to-day lives. These interfaces are designed to provide the most productive and user-friendly ways to interact with LLMs. They often also make it easy to include other sources of contextual information and chat personalization with ways to manage historical conversations and tap into external tools.

Each of these methods of interacting with LLMs provides different advantages and entails distinct challenges, making the choice dependent on your use case, customization requirements, technical expertise, resource availability, and budget constraints. Finally, in the next section, we'll briefly consider a new type of model that has been growing in popularity: the SLM.

The growth of the small language model (SLM)

When considering the complexities of making decisions around LLM technology and the slowdown in huge gains of the latest and greatest LLMs, it's worth introducing the SLM, which has been growing in popularity. SLMs are challenging the notion that bigger is always better and that size is not necessarily everything (pardon the pun). Model size might not be the sole determinant of performance, and factors such as architecture, training data, and fine-tuning techniques play a significant role.

Are LLMs reaching their limit?

As performance levels off, it begs the question: are LLMs reaching their limit? When ChatGPT was first released, we were all universally wowed by its vast knowledge, but subsequent iterations haven't been as revolutionary. When it comes to training data, improvements have been made as fresher training data has been used with new GPT versions, but in reality, there will always be limitations in LLM data unless a model is trained regularly, which seems unlikely.

Despite their impressive capabilities, LLMs come with significant drawbacks, which I've covered throughout this book. Here is a quick recap. They require an enormous amount of data and parameters for training, leading to high computational power and energy consumption. This results in steep costs, putting core LLM development out of reach for smaller organizations and individuals. Additionally, LLMs present a steep learning curve for developers due to the complexity of the tools and techniques involved, and the long cycle time, from training to deployment, can slow down experimentation and innovation efforts.

Other challenges include bias in training data as well as hallucination issues, slow inference performance, high token costs, and serious security implications if you are looking to consume and infer from sensitive data. The analogy of LLMs being like sledgehammers cracking nuts highlights that vast internet knowledge may not be essential for many conversational AI use cases. This raises the question: What qualities are truly important for effective conversational AI projects, and can an SLM meet these requirements and even provide advantages over LLMs?

Enter SLMs

Essentially, an SLM is a compact version of larger AI models, crafted to comprehend, interpret, and produce human language. SLMs offer a streamlined alternative to LLMs. With fewer parameters and simpler designs, they require less data, training time, and computing power – often just minutes or a few hours. This makes SLMs more efficient and easier to implement on-site or on smaller devices and with much smaller datasets fine-tuned to specific use cases.

SLMs versus LLMs – key differences

To gain a clearer insight into SLMs, it's helpful to compare them directly with LLMs:

- **Scale and scope**: SLMs are designed to be compact and efficient, making them suitable for specific domains and tasks. They are trained on smaller datasets, allowing for quicker training and inference times and targeted knowledge. LLMs, on the other hand, are larger and more comprehensive, trained on vast and diverse data sources. LLMs capture a broad range of language patterns and excel at generating highly coherent and contextually relevant text about practically any subject.

- **Training time and computational resources**: LLMs require more computational resources and longer training times due to their size. SLMs, with their smaller size and less complex architecture, are more practical for limited-resource scenarios and budgets or when quick deployment is needed

- **Domain expertise**: While both models can be fine-tuned to specific domains, SLMs are often more efficient for tasks requiring domain-specific expertise due to their smaller size and faster inference times. SLMs can also be trained on much smaller datasets. A well-trained domain-specific SLM may perform adequately enough when compared to an LLM with domain knowledge provided at the context level.

- **Versatility and costs**: It all depends on the use case, but LLMs shine in content generation, translation, and understanding complex queries due to their broader knowledge base. However, properly fine-tuned SLMs can achieve comparable performance for specific conversational tasks at a fraction of the computational cost. An SLM does not necessarily need to understand every subject. SLMs can be trained to interact with customers on the basis of specific knowledge of a particular domain, providing tailored solutions and improving user experiences. This is achievable by fine-tuning SLMs with specialized training data and can be more straightforward. Furthermore, fine-tuning SLMs demands few resources. The specific hardware requirements, and thus the associated costs, vary depending on the model size, complexity, and dataset needs, but these are likely to be less.

Advantages of SLMs

One of the key benefits of SLMs is their suitability for specific applications. Their focused scope and reduced data requirements make them ideal for fine-tuning particular domains or tasks. This customization enables companies to create SLMs tailored to their specific needs, such as sentiment analysis, **named entity recognition (NER)**, or domain-specific question-answering.

SLMs also offer enhanced privacy and security. Their smaller code base and simpler architecture make them easier to audit and less likely to contain vulnerabilities. This is especially attractive for sensitive data handling in industries such as healthcare and finance. The reduced computational requirements of SLMs also enable local processing, further improving data security.

SLMs are less prone to hallucinations within their specific domains. Their training on targeted datasets helps them learn relevant patterns, vocabulary, and information, reducing the likelihood of irrelevant or inconsistent outputs. With fewer parameters, SLMs are also less likely to capture and amplify noise or errors in training data.

Disadvantages of SLMs

SLMs, despite their numerous advantages, face several disadvantages and have been observed to have performance drops in some circumstances. A key limitation is their reduced understanding of context and overall less comprehensive knowledge base compared to larger models. This makes SLMs less effective for complex tasks that require deep understanding or broad contextual awareness. Additionally, SLMs, because of their smaller size and focused training datasets, may struggle with generating coherent and fluent text. For example, models with fewer parameters often show limitations in maintaining coherence over longer texts or in more complex interactions. This may constrain their utility in applications that demand high-quality responses while conversing over technical or complex domains.

Some examples of SLMs

Phi-3 is is the latest family of lightweight Mini 3B and 14B Medium state-of-the-art open models by Microsoft which get some really impressive performance results in benchmarks. Google's Gemma is a series of SLMs designed for efficiency and user-friendliness. As with other SLMs, Gemma models can run on everyday devices without specialized hardware. Cerule, a powerful image and language model, combines Gemma 2B with Google's **Sigmoid loss for Language-Image Pre-training** (**SigLIP**), leveraging efficient data selection techniques. This makes Cerule well suited for edge computing use cases. Another example is Llama 2 7B and now Llama 3 8B, which do not require vast hardware requirements, as well as Stable Beluga and Hugging Face's Zephyr. Many of these models are available on platforms such as Hugging Face and the LLM API providers, as well as cloud services such as Google Vertex AI. These models are all trained toward conversational use cases. A great way to try out these models is by using Ollama, a lightweight framework for running models on a local machine.

The transformative potential of SLMs

SLMs have the potential to democratize AI access and drive innovation. Their faster development cycles and improved efficiency enable cost-effective, targeted solutions. Deploying SLMs at the edge opens up possibilities for real-time, personalized, and secure applications across sectors such as finance, entertainment, automotive, education, e-commerce, and healthcare.

Edge computing with SLMs improves response times, data privacy, and user experiences by processing data locally, which is an appealing solution and one that rectifies many drawbacks for enterprises looking to employ LLMs in more sensitive areas. This decentralized approach to AI has the potential to transform how businesses and consumers interact with LLM technology.

By exposing SLMs to specialized training data and tailoring their capabilities with fine-tuning, they can produce accurate and relevant outputs for conversational use cases at a fraction of the cost of training an LLM. They often provide great results with similar conversational capabilities to an LLM. If you're looking to use LLM technology but are concerned about the drawbacks, it's advisable to consider whether a SLM model may be a sensible choice.

However, there are limitations: while SLMs offer significant benefits, especially in terms of efficiency and cost, their applications are sometimes restricted depending on the planned use case.

Despite any drawbacks, they make a compelling case. With the growing availability of model-specific **platforms as a service** (**PaaSs**) and model training platforms, it's becoming increasingly convenient to explore and test the capabilities of SLMs, making it well worth the time and effort to investigate.

Where to go from here

Throughout this book, we've focused on crafting engaging conversational experiences with ChatGPT and LLMs both as a tool and as technology for complex conversational experiences. The landscape of LLMs has undergone a dramatic transformation since we first set out on this journey, fueled by continuous advancements in LLM technology. ChatGPT, while a groundbreaking innovation at its release, now represents just one piece of a much larger and more versatile LLM ecosystem.

I've worked in conversational AI for a number of years, many years before ChatGPT, and had the pleasure of working on and putting into production many NLU-based applications. This has given me a strong understanding of the challenges of conversational AI, what LLMs need to do to overcome them, and whether they are capable. There are thousands of NLU-powered conversational AI applications working today, many of which you could argue could be replaced by ChatGPT applications. Or at least elements of them could be, which is often dependent on the underlying technologies of the existing platforms and how quickly and effectively vendors have built in the ability to leverage LLM technologies alongside traditional intent-based systems.

Like everyone, I was initially amazed by the release of ChatGPT and the endless possibilities of the technology. However, I've always tried to be realistic about the adoption of the technology and its limitations and dangers. It's all dependent on the use case, of course. If you want to chat in depth about your own data, then a RAG system is a huge advancement in conversational AI, but in a transactional role, it's wise to have some restraint.

I personally feel the biggest shift moving forward lies in the feasibility of building transactional LLM-powered applications for more complex use cases. Creating these applications with frameworks such as LangChain has become a much more realistic proposition but is still not by any means straightforward. With the emergence of supporting tools such as LangSmith, designed specifically for testing and monitoring LLM behavior, the risks of deploying transactional bots are mitigated. This opens doors to consider replacing traditional intent-based systems with the flexibility and adaptability of LLMs.

The future of LLMs lies in a multi-model approach. We can leverage different models suited to specific tasks within an LLM pipeline. Imagine a specific LLM acting as the agent executor within a LangChain application, drawing upon the strengths of other specialized models to complete specific tasks. Additionally, advancements in LLM reasoning capabilities offer exciting potential.

However, model choice remains a complex decision when planning for your LLM application. Factors such as use case, underlying data, domain type, cost, security considerations, and maintainability all play a crucial role when planning for production. These factors highlight the importance of exploring alternatives beyond ChatGPT. SLMs offer an intriguing option for specific scenarios, particularly with long-standing conversational AI systems, with a wealth of conversation logs and historical usage data that can be leveraged for fine-tuning.

The LangChain ecosystem remains a powerful platform for crafting LLM-powered applications. With integrations and tools tailored to practically any requirement, we've barely scratched the surface of its potential. You can leverage an ever-growing array of vector stores and services alongside specialized tools such as Tavily Search to enhance the capabilities of our LLM applications. There are many great examples of LangChain templates, which I advise are great places to look at more complex conversational use cases.

It's also important to acknowledge the documented risks involved in taking LLMs into production. Their stochastic nature, meaning inherent randomness in outputs, can lead to unreliable results. This necessitates careful planning and robust testing methodologies, especially for critical applications.

OpenAI continues to push boundaries with its GPT models. We've witnessed the evolution from GPT-3.5 to GPT-4 and its Turbo iteration, optimized for chat-based applications. The latest, GPT-4o released in May 2024, integrates multi-modal capabilities by incorporating their modality-specific models into one new model. The result is that it's capable of handling text, audio, image, and video inputs with human-like response speeds, enabling fully multi-modal conversational interactions. Previously, GPTs have been dogged by latencies of 3-6 seconds on average over voice interactions.

While OpenAI continues to enhance its models with breathtaking capabilities, multi-modal, multi-lingual, expanded context windows, improved accuracy and improved LLM reasoning, they have also wisely kept API GPT-4o costs down and included it in the Free version of ChatGPT to make this technology accessible to a wider audience. While OpenAI continues to enhance its models with breathtaking capabilities – multi-modal, multi-lingual, expanded context windows, improved accuracy, and improved LLM reasoning – they have also wisely kept API GPT-4o costs down and included it in the free version of ChatGPT to make this technology accessible to a wider audience.

The landscape, as we first knew it, has morphed dramatically, mirroring the rapid evolution of LLM technology. The LLM ecosystem continues to thrive, presenting developers with a growing number of models and simplified tools for building and leveraging their capabilities. As the technology continues to evolve, so too will the possibilities for crafting transformative conversational experiences. Embrace experimentation, explore the ever-expanding LLM ecosystem, and stay at the forefront of this exciting revolution.

Summary

In this chapter, we looked into crucial considerations and challenges associated with transitioning a ChatGPT-powered system from a POC to a fully operational production environment, highlighting the significant responsibilities and risks involved in deploying LLM-powered applications, particularly RAG systems. By now, you should have an understanding of the importance of robust safeguards to prevent inappropriate responses and ensure reliability. We reviewed an incident where a chatbot malfunctioned due to poor content moderation and insufficient system controls, highlighting the need for proper safeguards.

A substantial portion of the chapter was devoted to discussing specific challenges faced by RAG systems as they approach production. We also explored comprehensive strategies for evaluating LLM applications, emphasizing the necessity of continuous validation against a wide array of criteria to ensure the system's effectiveness and reliability. I'm hoping by covering an evaluation example, you now have a solid understanding of the importance of a tool such as LangSmith and a good base of knowledge to build robust evaluation. Evaluation and monitoring tools offer a structured approach to managing, testing, and refining LLM applications, and they need to play an important role in your ChatGPT project.

To give a balanced view, in the chapter, we gained some insights into alternative LLM service offerings and models, including alternatives to LangChain and ChatGPT, reflecting on the rapidly evolving landscape of conversational AI technologies. The landscape has moved on considerably, and you'll hopefully have a grasp of the importance of choosing the right model based on specific use cases and operational requirements. We also looked at the potential of SLMs as a more manageable and specialized alternative to traditional large-scale LLMs.

Finally, we looked at where to go from here, and that depends on where you and your organization are in your LLM journey, whether that be moving forward with implementing LLM features, transitioning your conversational AI projects into production, or embarking on greenfield LLM applications.

We've covered a lot here, but hopefully, it will help you overcome challenges and ask key questions so that you can successfully leverage LLM technologies moving forward.

Further reading

The following links are a curated list of resources to help in this chapter:

- LangChain: `https://chat.langchain.com`
- LangChain templates: `https://templates.langchain.com`
- LangSmith: `https://docs.smith.langchain.com`
- RAGAs: `https://github.com/explodinggradients/ragas`
- `https://docs.ragas.io/en/latest/`
- LlamaIndex: `https://www.llamaindex.ai`
- HuggingChat: `https://huggingface.co/chat/`
- ChromaDB: `https://docs.trychroma.com/`
- LLMs: `https://www.together.ai`
- Ollama: `https://ollama.com`
- Quiq: `https://quiq.com`

Index

packtpub.com

Subscribe to our online digital library for full access to over 7,000 books and videos, as well as industry leading tools to help you plan your personal development and advance your career. For more information, please visit our website.

Why subscribe?

- Spend less time learning and more time coding with practical eBooks and Videos from over 4,000 industry professionals

- Improve your learning with Skill Plans built especially for you

- Get a free eBook or video every month

- Fully searchable for easy access to vital information

- Copy and paste, print, and bookmark content

Did you know that Packt offers eBook versions of every book published, with PDF and ePub files available? You can upgrade to the eBook version at packtpub.com and as a print book customer, you are entitled to a discount on the eBook copy. Get in touch with us at customercare@packtpub.com for more details.

At www.packtpub.com, you can also read a collection of free technical articles, sign up for a range of free newsletters, and receive exclusive discounts and offers on Packt books and eBooks.

Other Books You May Enjoy

If you enjoyed this book, you may be interested in these other books by Packt:

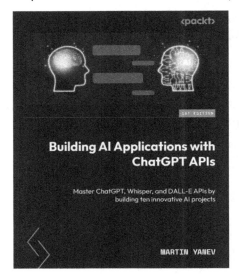

Building AI Applications with ChatGPT APIs

Martin Yanev

ISBN: 978-1-80512-756-7

- Develop a solid foundation in using the ChatGPT API for natural language processing tasks
- Build, deploy, and capitalize on a variety of desktop and SaaS AI applications
- Seamlessly integrate ChatGPT with established frameworks such as Flask, Django, and Microsoft Office APIs
- Channel your creativity by integrating DALL-E APIs to produce stunning AI-generated art within your desktop applications
- Experience the power of Whisper API's speech recognition and text-to-speech features
- Discover techniques to optimize ChatGPT models through the process of fine-tuning

The Future of Finance with ChatGPT and Power BI

James Bryant, Aloke Mukherjee

ISBN: 978-1-80512-334-7

- Dominate investing, trading, and reporting with ChatGPT's game-changing insights
- Master Power BI for dynamic financial visuals, custom dashboards, and impactful charts
- Apply AI and ChatGPT for advanced finance analysis and natural language processing (NLP) in news analysis
- Tap into ChatGPT for powerful market sentiment analysis to seize investment opportunities
- Unleash your financial analysis potential with data modeling, source connections, and Power BI integration
- Understand the importance of data security and adopt best practices for using ChatGPT and Power BI

Packt is searching for authors like you

If you're interested in becoming an author for Packt, please visit `authors.packtpub.com` and apply today. We have worked with thousands of developers and tech professionals, just like you, to help them share their insight with the global tech community. You can make a general application, apply for a specific hot topic that we are recruiting an author for, or submit your own idea.

Share Your Thoughts

Now you've finished *ChatGPT for Conversational AI and Chatbots*, we'd love to hear your thoughts! Scan the QR code below to go straight to the Amazon review page for this book and share your feedback or leave a review on the site that you purchased it from.

`https://packt.link/r/1-805-12953-8`

Your review is important to us and the tech community and will help us make sure we're delivering excellent quality content.

Download a free PDF copy of this book

Thanks for purchasing this book!

Do you like to read on the go but are unable to carry your print books everywhere?

Is your eBook purchase not compatible with the device of your choice?

Don't worry, now with every Packt book you get a DRM-free PDF version of that book at no cost.

Read anywhere, any place, on any device. Search, copy, and paste code from your favorite technical books directly into your application.

The perks don't stop there, you can get exclusive access to discounts, newsletters, and great free content in your inbox daily

Follow these simple steps to get the benefits:

1. Scan the QR code or visit the link below

https://packt.link/free-ebook/978-1-80512-953-0

2. Submit your proof of purchase

3. That's it! We'll send your free PDF and other benefits to your email directly

www.ingramcontent.com/pod-product-compliance
Lightning Source LLC
Chambersburg PA
CBHW080637060326
40690CB00021B/4966